THE
LITTLE BOOK OF
BIG
HISTORY

THE
LITTLE BOOK OF
BIG
HISTORY

The Story of Life, the Universe and Everything

IAN CROFTON
and
JEREMY BLACK

Michael O'Mara Books Limited

First published in Great Britain in 2016 by
Michael O'Mara Books Limited
9 Lion Yard
Tremadoc Road
London SW4 7NQ

A CIP catalogue record for this book is available from the British Library.

Papers used by Michael O'Mara Books Limited are natural, recyclable products made
from wood grown in sustainable forests. The manufacturing processes conform to the
environmental regulations of the country of origin.

ISBN: 978-1-78243-429-0 in hardback print format
ISBN: 978-1-78243-685-0 in paperback format
ISBN: 978-1-78243-430-6 in e-book format

2 3 4 5 6 7 8 9 10

Additional research: Claire Crofton and Chloe Evans
Jacket design by Ana Bjezancevic
Picture credits: Shutterstock

Designed and typeset by Design 23, London

Printed and bound by CPI Group (UK) Ltd, Croydon, CR0 4YY

www.mombooks.com

Contents

PART THREE: HUMANS START TO DOMINATE

PART FOUR: CIVILIZATION

PART FIVE: THE RISE OF THE WEST

PART SIX: THE MODERN WORLD

Note: Rather than the Christian BC and AD chronology, more recent dates in Earth's history appear here as BCE and CE, before and after the start of the Common Era.

PART ONE

SETTING THE SCENE

How did we get to where we are now? The back story to the chronicle of humanity is a long one. There would be no human history without a physical place for it to unfold. So to truly understand ourselves, we have to understand how the universe came into being, how the stars and planets formed, why our planet has the right conditions for life to have appeared. And we also need to understand how living things work, and how they evolved, and how we have ended up – with us.

TIMELINE

13.8 billion years ago: The Big Bang brings the universe into existence.

4.6 billion years ago: Formation of our solar system, including the Sun, the Earth and the other planets.

4.5 billion years ago: The Moon is formed, probably as a result of a collision between the Earth and a Mars-sized planet.

4.2 billion years ago: Oceans may have begun to form.

4.1–3.8 billion years ago: Earth and other inner planets suffer numerous impacts from asteroids.

4 billion years ago: Formation of oldest rocks still present on the Earth. Possible appearance in the oceans of self-replicating molecules, such as DNA.

3.7 billion years ago: Earliest indirect evidence of life on Earth suggests bacteria-like organisms feeding on organic molecules.

3.4 billion years ago: Cyanobacteria (blue-green algae) emerge, which draw energy from photosynthesis.

2.45 billion years ago: Start of the build-up of free oxygen in Earth's atmosphere, as a by-product of photosynthesis.

IN THE BEGINNING

Before the advent of modern science, there was a range of beliefs about the age of the Earth, and of the universe. Some Christians believed that God created both a mere 6,000 years ago. Ancient Hindu texts, in contrast, talk of an infinite cycle of creation and destruction.

Towards the end of the 18th century, geologists began to realize that the Earth must be much more ancient than had been thought (at least in Europe) – perhaps millions if not billions of years old. However, into the 20th century the scientific consensus was that the universe itself was eternal, and in a 'steady state'. Stars might be born and die, but the dimensions of the universe were fixed and unchanging.

A chink in this theory came in the 1920s when the American astronomer Edwin Hubble observed that the further away a galaxy is from us, the faster it is receding. He concluded that the universe is expanding, and that this expansion started in a single great explosion, which became known as 'the Big Bang'.

Arguments persisted between the proponents of the steady state and those of the Big Bang. Then in 1964 two radio astronomers working in New Jersey, Arno Penzias and Robert Wilson, noticed that their sensitive microwave receiver was suffering from constant interference, the same in all directions, with a wavelength representing a temperature of 2.7 degrees above absolute zero. At first they thought the phenomenon might be caused by the proximity of New York City or by pigeons defecating on their instrument. Eventually they realized that what their receiver was picking up was an echo of the Big Bang. If you retune your radio, part of the 'white noise' you hear between stations is this very same echo from the beginning of time.

9 billion years later, formation of the solar system and the Earth

300 million years later, beginning of the formation of stars and galaxies

380,000 years later, electrons and nuclei combine into atoms

The universe expands and cools

A few minutes after the Big Bang, formation of simple nuclei

One-millionth of a second after the Big Bang, formation of protons and neutrons

13.8 billion years ago, the Big Bang

The Big Bang

Cosmologists have now come up with a timetable that positions the Big Bang about 13.8 billion years ago, at a single point, a singularity, whose density and temperature were infinite. Once expansion started, it came at unimaginable speed. Between 10^{-36} and 10^{-32} seconds, the volume of the universe expanded by a factor of at least 10^{78}.* At this stage the only matter was elementary particles such as quarks and gluons. At about 10^{-6} seconds, as expansion slowed down and temperatures fell, quarks and gluons came together to form protons and neutrons. A few minutes later the temperature had cooled further, to about 1 billion degrees, and protons and neutrons combined to form the nuclei of deuterium and helium, though most protons remained unattached as hydrogen nuclei. Eventually, the positively charged nuclei attracted negatively charged electrons to create the first atoms. These simple atoms were to become the building blocks of the stars.

'Why does the universe go to all the bother of existing?'

Stephen Hawking, *A Brief History of Time* (1988)

THE BIRTH AND DEATH OF STARS

As the early universe expanded, matter was evenly distributed through space. But as tiny irregularities in density began to appear, gravity began to play a role, with denser regions attracting more and more matter. In this way clouds of gas, largely comprising hydrogen and helium, were formed. These so-called nebulae were where stars were – and continue to be – born.

Within a nebula, denser areas may begin to collapse in on themselves because of gravity, and these areas may eventually become dense and hot enough for nuclear fusion to begin – a reaction in which

---

* 10^{-36} = one divided by ten 36 times. 10^{78} = one multiplied by ten 78 times.

hydrogen is converted to helium, producing vast amounts of heat and light. It is this process that causes the stars – including the Sun – to shine with such intense brightness.

Just as gravity pulls together denser areas of gas to form stars, so it gathers stars to form galaxies. Our galaxy, the Milky Way, contains 100–400 billion stars and has a diameter of around 100,000 light years – meaning that light travelling at a speed of 300,000 kilometres per second takes 100,000 years to pass across it. Our Sun lies on one of the spiral arms of our galaxy, about 30,000 light years from the centre. The nearest star to the Sun is Proxima Centauri, just 4.24 light years away. The Milky Way is one of at least 100 billion galaxies in the universe. The size of the universe is a subject of speculation, but the part of it we can observe is 93 billion light years in diameter.

'The wonder is, not that the field of the stars is so vast, but that man has measured it.'

Anatole France, *The Garden of Epicurus* (1894)

Different sizes of stars may undergo particular sequences in their lifetimes. Those similar in size to the Sun burn at something like 6,000 degrees on the surface (the core is much hotter) for at least 10 billion years before they exhaust their hydrogen. At this stage, the core contracts and the temperature rises to 100 million degrees, allowing helium fusion to begin. The star expands to become a red giant, around 100 times larger than in its youth, before shrinking to become a white dwarf, 100 times smaller than the original.

Larger stars have shorter lives. For example, a star ten times the size of the Sun will turn into a red giant after only 20 million years. As the temperature increases, the star begins to synthesize heavier and heavier elements, until at 700 million degrees iron is created. This process is the origin of many of the elements that make up planets such as the Earth –

not only iron, but also carbon, oxygen and silicon. At this point the star blows apart in a massive explosion called a supernova, a fast-expanding cloud of gas and dust. At its centre is an object called a neutron star, only 10 to 20 kilometres in diameter, but so dense that a cubic centimetre of its material has a mass of 250 million tonnes. Even larger stars may end their lives as a black hole, an area of space so dense that not even light can escape its immense gravitational pull. There may be a supermassive black hole at the centre of our own galaxy.

THE GOLDILOCKS ZONE

The solar system – the Sun and its planets – formed about 4.6 billion years ago from a nebula – a spinning cloud of dust and gas. As denser patches of dust attracted more and more material by force of gravity, so the planets were formed. They all still spin in the same direction.

Earth is less than one-tenth of the size of the Sun's largest planet, Jupiter, and Jupiter only one-tenth the size of the Sun. The Earth is 149,600,000 km from the Sun, Jupiter is five times further out, and the outermost major planet, Neptune, thirty times further. The relatively small inner planets – Mercury, Venus, Earth and Mars – are rocky in composition, whereas the giant outer planets – Jupiter, Saturn, Uranus and Neptune – mostly consist of gas surrounding a small rocky core.

Life as we know it is based on the cell, and for cells to function water must exist in a liquid state. Both Mercury and Venus are too close to the Sun for this to happen. It is possible that the conditions for life might once have existed on Mars, and NASA's rovers on the surface of the planet are exploring this possibility. The outer planets are much too cold to support life, although liquid water may exist under the surface of some of their moons.

As far as we know, though, Earth is the only planet in the solar system that houses life. Earth is said to lie in the 'Goldilocks zone', the region around a star where the conditions are just right for life. In the tale of Goldilocks and the Three Bears, Goldilocks picks the porridge that is neither too hot nor too cold, the chair that is neither too small nor too big, and the bed that is neither too hard nor too soft. Earth is neither too close nor too far away from the Sun (and thus not too hot nor too cold) for water to exist as a liquid. It is large enough to generate a strong gravitational field to hold on to an atmosphere, and thus has sufficient atmospheric pressure to allow liquid water to exist on the surface.

Are we alone in the universe?

Recent detailed observations of our own galaxy suggest that it may contain as many as eleven billion Earth-size planets orbiting Sun-like stars within the Goldilocks zone. It is thought that the nearest such planet is twelve light years away, meaning that it would take twelve years for a radio signal twelvem Earth to reach it. But having these minitwelveal conditions does not necessarily mean that a planet does possess life – let alone a form that has evolved enough to send us a radio signal. Indeed, although radio telescopes around the world have been monitoring the airwaves for decades, no signs of intelligent extraterrestrial life have been detected.

THE RESTLESS EARTH

Our planet is a not-quite-regular sphere, layered like an onion. In the centre, its inner core consists of solid iron. Around this lies first the outer core, of molten iron, and then the mantle, made up of molten rock called magma. Floating on top of the mantle is a thin crust made of solid rock. We live on the surface of the crust. Although humans have been to the Moon, no one has gone deeper below the surface than 4 km, the depth of the deepest mine.

The Earth has one more layer, a gaseous skin. This is the atmosphere, more than three-quarters of which is nitrogen and one-fifth oxygen, essential to most forms of life. There are small amounts of other gases, but of these carbon dioxide and methane – the so-called greenhouse gases – have a crucial bearing on life on Earth (see p. 246), as does the presence of water vapour, an essential component in all weather systems. The density of the atmosphere grows thinner with altitude and gradually fades into space.

Just as the gases in the atmosphere are constantly in motion, so too are the rocky plates that make up the crust. Scientists used to assume that the continents and seas had always been in the same positions. Then in 1915 a German meteorologist called Alfred Wegener suggested that rather than being static, the continents had drifted over time. He had observed that the rocks and fossils along the east coast of South America were similar to those on the west coast of Africa, and that certain extinct plants were found not only in these two locations, but also in Madagascar, India and Australia.

Over the years, more and more evidence came to light to support Wegener's theory of continental drift. It became clear that this process had had a crucial impact on the distribution and dispersal of different groups

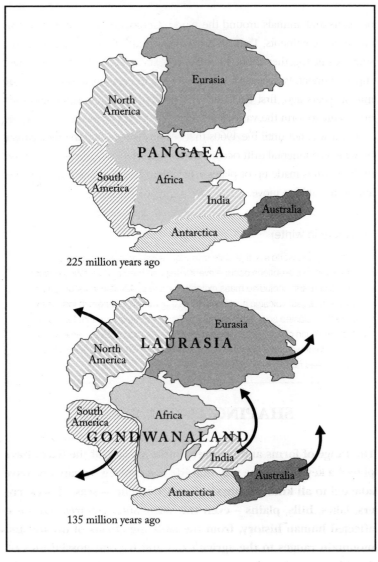

Eurasia

North
America

PANGAEA

South
America

Africa

India

Australia

Antarctica

225 million years ago

Eurasia

North
America

LAURASIA

South
America

Africa

GONDWANALAND

India

Australia

Antarctica

135 million years ago

Continental drift

of plants and animals around the world. Geologists now agree that two enormous continents, Laurasia in the north and Gondwanaland in the south, came together about 300 million years ago to form an even bigger supercontinent, Pangaea. This in turn began to break up about 200 to 180 million years ago, first back into the original two continents, and then eventually to form the various separate continents of today.

But it was not until the 1960s that scientists identified the mechanism by which continental drift occurs, and named it plate tectonics. The crust of the Earth is made up of plates which float on top of the liquid mantle, and so are able to move.

Volcanic winter

It is along the world's active plate boundaries that most earthquakes and most volcanic eruptions occur – events that can have a devastating impact on life on Earth, including mass extinctions (see p. 42). Within recorded history, the largest volcanic eruption was that of Mount Tambora in Indonesia in 1815. It blasted so much ash into the Earth's atmosphere that for many months much of the Sun's light was blocked out, and 1816 became known as 'the year without a summer'. Crops failed and livestock died, resulting in widespread famine in Europe and North America.

SHAPING THE SURFACE

The range of forms and features on the surface of the Earth have played a key role in the way that life has evolved. Organisms have adapted to all kinds of physical environments – seas, shores, rivers, lakes, hills, plains – even the skies. Such features have also affected human history, from the isolating effects of oceans and mountain ranges to the agricultural and trading possibilities offered by great rivers.

The fundamental building material of the Earth's surface is rock. We perceive it as solid and enduring, but over aeons it can be destroyed and recreated. The series of processes involved is called the rock cycle, and is powered partly by the Sun and partly by the heat below the Earth's crust.

The Sun's heat causes water to evaporate. This forms clouds, which precipitate as rain or snow. Water erodes rock, ice splits it, and snow builds up into glaciers, which grind away at the rock as they flow downhill. Rivers wash away the eroded material, which is deposited elsewhere, as clay or sand, usually at the bottom of seas. As layers of such sediments build up they are compressed into rock. Some deep sedimentary rocks experience so much pressure from above and so much heat from below that over long periods of time they metamorphose into entirely different types of rock. Quartzite, for example, is metamorphosed sandstone. The third kind of rock, in addition to sedimentary and metamorphic, is igneous. This is formed as magma deep beneath the crust rises towards the surface. Sometimes the magma is trapped beneath the surface, forming rocks such as granite. Sometimes it finds its way, via volcanoes and fissures, to the Earth's surface, where it solidifies into rocks such as basalt.

The movement of the Earth's tectonic plates also plays its part. Where one plate is pushed deep under another, its rocks are absorbed into the molten mantle below. Where two plates are pulling apart from each other, as happens in the middle of oceans, molten rock comes to the surface to form great mid-ocean ridges. Volcanic activity such as this has also created – and destroyed – mountains elsewhere. Mountain ranges may also be formed by one plate pushing against another, which folds up the previously horizontal sedimentary layers. In this way the collision of India with the rest of Asia formed the Himalayas, which are still rising in height by around 1 centimetre per year.

The shapes of the landscape can be altered by other processes. Both rivers and glaciers carve out valleys, and rivers can create new areas of coastal land in the form of deltas, formed from sediments. Ocean currents and wave action also alter coastlines, eroding material and depositing it elsewhere. Such changes can have important human impacts. The creation of a delta provides rich ground for farming, for example, while those who depend on the sea for their livelihood may be left high and dry by a receding coastline.

WHAT IS LIFE?

All living things react to stimuli, feed, grow, reproduce, repair themselves and die. Some of these attributes are found in non-living things – crystals 'feed' on salts dissolved in water, and thus grow; robots can respond to stimuli. So what sets living things apart?

The answer is the cell, the basic unit of life as we know it. An individual cell is tiny – the smallest fraction of a millimetre in diameter. But cells are among the most complex mechanisms known to science. Some are separate living organisms in their own right (see p.25), while others play specialized roles in more complex multicellular organisms (see p. 26). There may be as many as 37 trillion cells in a human body.

Cells have the ability to absorb a wide range of raw materials from their environment and alter them chemically within themselves to create more complex compounds. It is this ability that enables them to repair damage, and to reproduce by dividing and redividing.

Four groups of chemicals are essential for the structure and functioning of a cell. Nucleic acids (DNA and RNA) encode genetic information and carry out the instructions embedded in that code (see p. 26). The second group consists of proteins, some of which are structural, while others are enzymes – catalysts that help drive chemical reactions. Proteins are made from simpler building blocks, the amino acids. The third group is the carbohydrates, some of which are building blocks while others store energy. The simplest carbohydrate is glucose, produced by plants during photosynthesis. Nearly all animals acquire the carbohydrates they need from plants. The final group is the lipids, key components of the membrane of the cell (see p. 25).

All these complex molecules are made up of a relatively small range of simpler molecules, most notably water and carbon. Water contributes oxygen and hydrogen to many other compounds. In addition, some two-thirds of a living cell consists of liquid water, in which more complex compounds are dissolved and transported. Carbon has the ability to combine with other elements to create a huge variety of organic compounds, many of which are soluble in water.

The origin of life?

How did simple molecules such as water and carbon come to form the more complex compounds necessary for life? The atmosphere of the early Earth consisted of gases emitted during volcanic eruptions, such as water vapour (H_2O), hydrogen (H_2), nitrogen (N_2), carbon dioxide (CO_2) and carbon monoxide (CO). As these gases cooled, hydrogen combined with nitrogen to form ammonia (NH_3) and with carbon dioxide and carbon monoxide to form methane (CH_4). In the presence of ultraviolet light (as emitted by the Sun) and an electrical spark (e.g. a lightning flash), ammonia and methane can combine with water and carbon dioxide to form simple amino acids, which, if heated, can link up to form proteins. Similar reactions can result in the creation of the components of DNA. Alternatively, there is some evidence that not only the components of DNA but also the amino acids may have been brought to Earth by meteorites.

As the Earth cooled, water vapour in the atmosphere condensed to form the early oceans, in which many different minerals and gases were dissolved. It is possible that in this great chemical soup about 4 billion years ago the first self-replicating molecules – such as DNA – might have appeared.

WHERE DOES THE ENERGY COME FROM?

Virtually all the energy on which life on Earth depends, and ultimately much of the energy that humans use in modern industrial societies, comes from the Sun.

The key process in the natural world is photosynthesis, a series of chemical reactions by which the energy of sunlight is used to convert water and carbon dioxide into glucose, a simple carbohydrate that living organisms can use as a source of energy.

Life without the Sun?

Not all of life on Earth depends on energy from the Sun. Some energy comes from the layer of molten rock beneath the Earth's crust, via volcanic vents. In places at the bottom of the oceans, these so-called hydrothermal vents release hot water full of hydrogen sulphide. This gas is poisonous to most organisms, but provides a source of energy for certain bacteria. These form the base of strange communities including clams, limpets, shrimps and giant tube worms.

Those organisms that can make their own food in this way are known as primary producers. On land, most primary production is carried out by green plants. In the oceans, the organisms responsible for most primary production are the phytoplankton – microscopic single-celled organisms such as algae and diatoms.

These photosynthesizing organisms are at the bottom of all food chains. Primary producers are fed on by primary consumers – the herbivores. They in turn are fed on by secondary consumers – the carnivores. Sometimes even the carnivores may be preyed upon by top carnivores –

for example, a small insect-eating bird provides food for a hawk.

Because the laws of physics dictate that energy transfers are always inefficient, there tend to be fewer individuals in each level as one climbs the food chain. A herbivore typically only gains about 10 per cent of the energy available in the plant it eats. The rest is wasted in undigested material, and in heat loss through respiration.

The final stages in the energy pathway involve scavengers and decomposers. Scavengers such as woodlice and millipedes feed on the faeces and dead remains of plants and animals. Decomposers such as certain fungi and bacteria complete the process by using any energy left in the last bits of dead matter.

Humans have fitted into food chains in a range of different ways. Some communities have been mainly herbivores, gathering seeds, nuts and berries, or cultivating crops. Others have been primarily carnivores, hunting or scavenging their food, or rearing livestock such as cattle. But most communities, both past and present, have tended to be omnivorous, eating both animal and plant matter. Although we tend to think of ourselves as top of the food chain, in some ecosystems we have found ourselves on the menu for even bigger, more powerful carnivores.

For our energy needs beyond just food, we have in the past relied completely on the Sun. Firewood is plant matter, and fossil fuels – coal, oil and natural gas – all derive from plant matter. The power that we can extract from running water, waves, and wind comes from systems in the atmosphere driven by the Sun. Tidal power is rather different, relying on the gravitational pull of the Moon, and only to a much lesser extent that of the Sun. Geothermal energy exploits heat from deep beneath the Earth's surface, while nuclear power unleashes the energy locked within the nucleus of the atom.

LIFE GETS COMPLICATED

About 4 billion years ago, in the rich chemical soup of the early oceans, life may have begun to evolve. The key moment, scientists believe, would have been the appearance of complex organic molecules (like DNA) that were capable of replicating themselves.

At that point in Earth's history there was no ozone layer in the upper atmosphere to keep out the Sun's intense ultraviolet radiation. As these complex molecules replicated themselves, the solar radiation would have caused frequent mutations. Some of these could have yielded molecules that were better adapted to their environment than others. In this way natural selection may have begun.

For example, molecules that replicated more often and more accurately would hold an advantage, as would those that could use other molecules to build a protective layer. Experiments show that in conditions similar to those found in periods of fierce volcanic activity, followed by rapid cooling by cold water, amino acids can form into structures surrounded by a membrane. In such ways the first cells might have come into being.

These primitive cells would have been simple bacteria-like organisms known as 'prokaryotes', consisting of an outer membrane enclosing protoplasm, a gel-like material containing a range of small and large molecules. The DNA is located in a particular area of the protoplasm, but otherwise there is little in the way of structure. Prokaryotic cells reproduce by splitting into two new cells. These early microorganisms fed on the rich soup of organic molecules found in the early oceans.

Around 3.4 billion years ago, as the supply of organic molecules started to run out, a new group of prokaryotic microorganisms evolved. These were the cyanobacteria, and they had an alternative way of feeding: photosynthesis. Photosynthesis uses the energy from sunlight to convert carbon dioxide and water into glucose (a simple sugar), with oxygen as the by-product. Before this, oxygen had been poisonous to living things. Now, as oxygen built up in the atmosphere, many forms of life started to depend upon it.

The next big leap came 1.8 billion years ago, when larger, more complex cells appeared. These so-called eukaryotic cells contain the DNA within a central structure, the nucleus. There are also a number of other specialist structures with particular functions. These are called organelles. The fact that some of them have their own DNA, together with the resemblance between certain organelles and certain bacteria, led the US biologist Lynn Margulis to conclude in the late 1960s that eukaryotic cells started as symbiotic (mutually beneficial) associations between various types of prokaryotic cells. This theory is now generally accepted by scientists.

'Far from leaving microorganisms behind on an evolutionary "ladder", we are both surrounded by them and composed of them.'

Lynn Margulis and Dorion Sagan, *Microcosmos* (1986)

Although our own cells and those of all other multicellular organisms are eukaryotic, the first eukaryotic cells were single-celled microorganisms similar to many types that still exist, such as protozoans (which have some animal-like characteristics), slime moulds (which have some fungi-like features) and certain algae (which, like plants, photosynthesize).

Perhaps the most significant innovation that came along with eukaryotic cells was the beginning of sex (see p. 27). Sexual reproduction, in which some genetic elements come from one parent, some from the other, led to far greater variation. This in turn quickened the pace of evolution, as greater variation allows for faster adaptation to the changing environment.

HOW LIFE CARRIES ON

One of the defining characteristics of living organisms is their ability to reproduce. All things, both living and non-living, are subject to decay, so reproduction provides a means by which life can perpetuate itself. It's as close as we can come to immortality.

Simple single-celled organisms such as bacteria reproduce by splitting down the middle: one 'mother' cell becomes two 'daughter' cells. This is asexual reproduction. Barring random mutations, the mother and daughter cells are genetically identical. The cells that make up our own bodies reproduce in this way too, enabling us to repair damage, and to grow.

Multicellular organisms, including plants and some animals, can also reproduce asexually. In so-called vegetative reproduction, a new plant, genetically identical to the 'parent' plant, can grow from a bit of root, or a runner, or a leaf, or a twig. Some invertebrate animals, such as sea anemones, sponges and many marine worms, can reproduce asexually by 'budding', when a small part of the parent body grows, then separates to form a new individual.

In sexual reproduction, the genetic material from two specialized sex cells (sperm from the father, egg from the mother) combines to form a single new cell that inherits genetic characteristics from both. This new cell divides and divides, eventually developing into a new individual with a new and unique genetic make-up.

Although plants can reproduce asexually, they can also reproduce sexually. In flowering plants, the male sex cells are in the pollen, and this is transferred from one flower to another either by wind or by an animal such as a bee. Once the pollen lands on another flower, the female sex cell can be fertilized, and grows into a seed from which a new individual can grow. Basically the same process occurs in animals, although animals use a variety of methods to achieve fertilization. In fish, for example, the female lays her eggs in the water and the male then sprays his sperm on the eggs. In placental mammals, the male inserts his penis into the female's vagina and ejaculates his sperm, which finds its way to the female egg. The resulting embryo develops inside the mother's womb until birth.

Different animals have adopted different childcare strategies. Many aquatic creatures such as fish lay vast numbers of eggs, but thereafter do not care for their young. The result is that most are eaten by predators long before they can breed, but one or two might survive. At the other end of the spectrum, apes (including humans) usually bear only one or two offspring at a single birth, and spend many years rearing their children, who need care until adulthood.

THE ORIGIN OF SPECIES

The secrets of the evolution of life on Earth were long locked up in the planet's rocks. For centuries, drawing on the account of creation in the Bible, people had assumed not only that the Earth was very young, but that all species had remained unchanged since the beginning.

Then in the later 18th century the Scottish geologist James Hutton established that it must have taken millions of years for natural processes such as heat and erosion to make the geological landscape we know today. It followed that if the rocks were that old, so too were the fossils they contained.

Some of these fossils were completely unlike any creatures we know today, while others were similar, but different. By the start of the 19th century, naturalists were identifying what they thought of as a hierarchical progression, from the simplest life forms in the earliest rocks to more complex forms nearer our own time. One might call this progression 'evolution', but no one could explain why simpler life forms, such as bacteria and sponges, are still alive today. Nor could they explain the great variety of species that now populate the world. If evolution had occurred, how did it work?

It turned out that it worked blindly, without purpose or direction. It took the genius of Charles Darwin to identify this, and to discover the key mechanism: natural selection. As a young man in the 1830s Darwin had sailed as ship's naturalist on HMS *Beagle* on a long voyage of discovery, during which he noted the resemblances between creatures living on different continents, such as the flightless South American rhea and the African ostrich. He also noted how, for example, different finches on the Galapagos Islands had different beaks enabling them to exploit different food sources. It seemed likely that the rhea and the ostrich had an ancestor in common; the same was true of the Galapagos finches.

Darwin knew that his theory flew in the face of Christian orthodoxy. Humans were no longer set apart from other animals. Rather than being created in God's image, they were descended from ape-like ancestors. So Darwin bided his time, and accumulated evidence, before finally publishing *On the Origin of Species* in 1859.

Species evolve over time, Darwin proposed, because now and then, at random, an individual appears with a feature that better equips them to survive and reproduce than their fellow creatures. Over the generations those individuals with the favourable characteristic are more likely to survive, and so to pass on this characteristic to their offspring, than those without it. Thus species change, and become better able to adapt to new environments. This process was subsequently dubbed 'the survival of the fittest'.

'Man with all his noble qualities . . . still bears in his bodily frame the indelible stamp of his lowly origin.'

Charles Darwin, *The Descent of Man* (1871)

Some aspects of Darwin's theory have been modified over the years, but more and more evidence has built up that makes natural selection incontrovertible, including similarities in body plans and embryonic development between species, the existence of vestigial structures such as the human coccyx (a vestige of our ancestors' tails), and above all the evidence from DNA that enables us to compare our own genomes with those of other very different animals, and trace what we share and what we do not. It was the understanding of genetics and the role played in it by DNA that explained how characteristics arise and are passed on to the next generation in the first place.

THE BLUEPRINT OF LIFE

Darwin's theory of natural selection explained how new species come about. But Darwin did not know how parents pass on their characteristics to their offspring. Nor did he know how new characteristics arise. We now know many of the answers.

The units of heredity are genes. They determine everything from eye colour to increased risk of suffering certain diseases. Some inherited characteristics (such as the colour of rats) are determined by a single gene, but most (such as human height, weight and eye colour) are determined by a number of different genes. Genes are packaged in long strings of molecules called chromosomes, and in all cells except sex cells there are two sets of chromosomes.

What was unknown was how the instructions for replicating any given trait in the offspring were coded within the gene. In the 1940s scientists began to suspect that a very large and complex molecule called deoxyribonucleic acid (DNA) was involved. Then in 1953 the American James Watson and the Englishman Francis Crick, working at Cambridge University, announced they had worked out how DNA encodes genetic information.

'A structure this pretty just had to exist.'

James Watson, *The Double Helix* (1968), on the DNA molecule

Genetic code, they showed, is embedded in the structure of DNA. The DNA molecule is a double helix – two strands that spiral around each other. Each strand is a sugar-phosphate backbone, and each is joined to the other by sequences of pairs of just four chemical components called nucleotide bases. Each base pairs only with one of the other three bases. This structure explains how DNA can replicate itself, by the separation of the two strands, and so pass on genetic information to offspring.

The structure of DNA also explains how the code is embedded. Each sequence of three nucleotide bases (a codon) contains the instructions for the creation of a particular amino acid. Amino acids are the building blocks of proteins, crucial components of all cells (see p. 21). A gene consists of a sequence of codons coding for a single protein, followed by a stop codon. Some parts of the DNA do not themselves code for amino acids but are control centres for turning genes or groups of genes on and off.

The way that DNA works also explains how mutation leads to new characteristics – the key driving force in natural selection. A mutation is a change in the sequence of nucleotide bases as the DNA replicates itself. This occurs naturally, but exposure to chemicals or radiation can accelerate the mutation rate. Only those mutations that occur in eggs and sperm will be passed on to offspring, and only these matter for evolution. Many mutations are neutral, but some can damage offspring, while a few can be beneficial. Mutations in control genes may have a particularly big effect on the organism. It is the beneficial mutations that better equip an organism for its environment, and it is these that are likely to be passed on to future generations.

PART TWO

ANIMAL PLANET

The first simple animals appeared on Earth more than half a billion years ago. Over hundreds of millions of years, a great variety of creatures adopted a range of body plans and life styles. Some of the earliest animals, such as starfishes and sea urchins, proved successful and have survived. Others, such as the dinosaurs, dominated life on Earth for 165 million years, before becoming extinct. Modern humans emerged a mere 200,000 years ago, and have only dominated the planet for a fraction of that time.

TIMELINE

600 million years ago: The emergence of the first multicellular organisms.

542–488 million years ago: Cambrian period. Evolution of external skeletons leads to a great diversity of animal body plans, among them trilobites and brachiopods. First vertebrates, equipped with a notochord, precursor of the spinal column.

488–444 million years ago: Ordovician period. Great diversification of trilobites, lamp shells, gastropods and graptolites. Emergence of sea urchins, starfishes and ammonites. The end of the period sees the first evidence of land plants, and a mass extinction of many species.

444–416 million years ago: Silurian period. New forms of marine life follow the mass extinction, including scorpion-like animals and jawed fishes (at first cartilaginous, latterly bony). The first invertebrates, scorpions and wingless insects, appear on land, as do vascular plants such as club mosses.

416–359 million years ago: Devonian period. Huge coral reefs. First ferns. Emergence of primitive amphibians, the first four-legged animals, which begin to colonize land.

359–299 million years ago: Carboniferous period. First flying insects and first reptiles. Great proliferation of land plants, including conifers, which over time form extensive coal deposits.

299–251 million years ago: Permian period. Reptiles diversify. Period ends with mass extinction of many groups of marine animals, including trilobites. Many terrestrial groups also wiped out, making way for dinosaurs.

251–200 million years ago: Triassic period. Emergence of the dinosaurs on land. The first small mammals also appear.

200–145 million years ago: Jurassic period. Great diversification of the dinosaurs, turtles and crocodiles. Tropical forests. One of the first bird fossils, *Archaeopteryx*, appears towards the end of the period.

145–66 million years ago: Cretaceous period. Flowering plants emerge and begin to dominate on land. Grasses appear. The end of the period witnesses the sudden mass extinction not only of the dinosaurs, but also of the ammonites, ichthyosaurs and pterosaurs. Birds (descended from one group of dinosaurs) and mammals survive.

66–56 million years ago: Palaeocene epoch. Many new groups of mammals appear, including the first primates.

56–34 million years ago: Eocene epoch. Spread of mammals, including elephants, whales, rodents, carnivores and hoofed mammals.

34–23 million years ago: Oligocene epoch. Spread of grasslands, and first appearance of monkeys.

23–5.3 million years ago: Miocene epoch. Spread of horses, first appearance of apes. Many animals, such as frogs, snakes and rats, are very similar to those of today.

7 million years ago: Split between our ancestors and the ancestors of chimps and bonobos.

6 million years ago: Early humans begin to walk some of the time on their hind legs.

5.3–2.6 million years ago: Pliocene epoch. Origin of mammoths. Walking upright becomes the norm for early humans.

2.6 million years ago: Earliest evidence of human tool use.

2.6 million–11,700 years ago: Pleistocene epoch. Period of ice ages and warmer interglacial periods.

2.4 million years ago: Appearance of *Homo habilis*.

1.9 million–143,000 years ago: Dominance of *Homo erectus*.

200,000 years ago: Emergence of *Homo sapiens* (modern humans) in Africa, where they remain for another 100,000 or more years.

11,700 years ago–present: Holocene epoch. After the end of the last ice age, many large land animals such as mammoths become extinct. Humans come to dominate the planet.

THE FIRST ANIMALS

For eons, although Earth's oceans teemed with life, you would not have been able to see a single individual organism. That is because for billions of years the only living things consisted of no more than a single cell.

It is possible that some cells amassed into colonies, but the first fossil evidence of these appears only about 600 million years ago. These colonies were probably rather like sponges, the most 'primitive' animal still living, found in many parts of the oceans. In sponges, each cell can live independently, but can also function together with other cells. If a living sponge is broken up into fragments they will in time come together to form a new colony. Sponges are fixed to rocks, and feed on minute particles in the water. There is some coordination between cells, but no real nervous system.

By 590 million years ago a whole range of higher animals had emerged, with more clearly defined body plans and identifiable nervous systems. These creatures were still confined to the oceans, and included coelenterates (such as jellyfish and sea anemones), annelid worms and arthropods. Key characteristics of arthropods include bilateral symmetry, segmented bodies, multiple legs, eyes, and an exoskeleton – a hard external layer that protects the internal organs. All of today's arthropods – including crustaceans, spiders, scorpions and insects – are descended from some of these early creatures. Other arthropod groups, such as the trilobites, are long extinct.

The trilobites appeared during the Cambrian period (542–488 million years ago), which saw an 'explosion' of new animal types, including most invertebrate groups known today. Various possible explanations for the Cambrian explosion have been put forward. The increase in photosynthesizing organisms had generated more oxygen in the atmosphere. By this time the ozone layer had formed, protecting life on Earth from the Sun's ultraviolet rays. There was also a sudden increase in the oceans of calcium (a key component of hard body parts, such as exoskeletons) due to greater volcanic activity along the mid-ocean ridges. Ecological and evolutionary explanations suggest that an accelerated arms race between predator and prey, perhaps initiated by the development of the first primitive eyes, may have allowed prey to be detected by a predator at a distance – and vice versa.

Prototypes

During the Cambrian explosion a number of groups appeared that resemble no animals we know today. Among the bizarre forms found in the Burgess Shale of Canada was *Opabinia*, a predator with five eyes and a snout like a vacuum cleaner. Another predator was *Anomalocaris*, 60 centimetres long, propelled by lateral flaps, and with a mouth that resembled a slice of pineapple. Many of the Burgess designs were only brief Cambrian successes – but *Pikaia*, an eel-like creature about 4 centimetres long, has been interpreted as an ancestor of the vertebrates, members of the phylum Chordata, which includes humans.

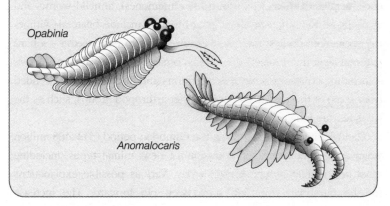

Opabinia

Anomalocaris

The next key event was the appearance of the first true vertebrates. Vertebrates are creatures with internal skeletons, including a spine made up of linked vertebrae protecting a spinal cord – the central component of more advanced nervous systems. A few obscure creatures living today have a spinal cord but not a spine, and it is thought that their ancestors first emerged in the Cambrian. But the first true vertebrates, the jawless fishes (similar to modern lampreys), appeared around 500 million years ago.

In these early fishes, the internal skeleton was made of cartilage, not bone. This is still the case with a number of fishes today, including sharks and rays, whose ancestors were the first to develop jaws, around 410 million years ago. Jaws are a characteristic of nearly all vertebrate groups today.

LIFE COMES ASHORE

Even after the appearance of large multicellular animals and plants, life remained confined to the oceans. One of the most diverse animal groups was the fishes. They have sets of highly specialized structures: gills that extract free oxygen from water, and fins to propel them through their element.

Gills and fins do not work on land, so when and how did the vertebrates crawl up onto the shores of the ancient continents? During the Silurian period (444–416 million years ago), the first invertebrates – scorpions and wingless insects – arrived on land. They offered a potentially rich food source for larger predators, and this may have lured the first vertebrate, an amphibian, on to land about 370 million years ago.

Amphibians are tetrapods (four-limbed animals), a body form they share with reptiles, birds and mammals. Quite when the tetrapods first evolved, and from which group, is the subject of debate. Some scientists trace them to lobe-finned fishes about 395 million years ago, others to lungfish. Both lobe-finned fish (which today survive only as two species of coelacanth) and lungfish appear to 'walk' on their strong bony fins under water. Their ability to breathe air enables them to survive the tropical dry season, when they burrow into mud and enter a period of dormancy.

Amphibians (which today include frogs, toads, newts and salamanders) divide their time between land and water. Their early stages – as eggs and then tadpoles – are entirely aquatic. Tadpoles have gills, but as they mature their lungs take over, enabling them to breathe air. The early amphibians probably spent most of their time in water, but, as they evolved stronger skeletons, more efficient lungs and skins less prone to dehydration, they spent more time on land.

For millions of years, large amphibians up to several metres long were the top land predators, occupying an ecological niche similar to that of modern crocodiles. But with the appearance of the first reptiles, better adapted to life on land, their days as top dog were numbered.

The first land plants

Primitive plants such as algae (including seaweeds) had thrived in the oceans for hundreds of millions of years. In the benign environment of the oceans, all parts of their surface could be involved in the gaseous exchange required for photosynthesis. The key feature of most groups of land plants, a vascular system to circulate water, nutrients and vital chemicals, had appeared by the end of the Silurian period, 416 million years ago. This system also gives structural support, crucial in competing to outgrow neighbours and so catch the maximum sunlight. Older growth accumulated as wood, and during the Carboniferous period (359–299 million years ago) forests of ferns, club mosses and horsetails grew as tall as modern trees. All the earliest plants reproduced by means of spores. The first seed-bearing plants, such as conifers, appeared towards the end of the Carboniferous period. Seeds, unlike spores, provide young plants with food, water and protection. Soon the seed-bearing plants came to dominate the land.

THE AGE OF THE DINOSAURS

Amphibians have never entirely adapted to life on land. Their dependence on water has limited the range of habitats they can exploit. It was not until the evolution of the reptiles that the first truly terrestrial vertebrates appeared.

Two factors explain the success of the reptiles – which include crocodiles, turtles, snakes, lizards and the extinct dinosaurs. The first is their ability to conserve water within their bodies. The second is the nature of the reptilian egg, which means they do not need to lay their eggs in water.

Inside a reptile's egg the embryo has its own watery environment, surrounded by a membrane. The yolk provides food for the embryo, while a further membrane allows for respiration and the excretion of waste matter. The albumen or egg white gives extra cushioning, and provides water and proteins. All this is surrounded by another protective membrane and the shell.

The first reptiles appeared about 340 million years ago, during the Carboniferous period: small creatures, only about 20 centimetres long, that lived in that period's lushly vegetated swamps – the perfect habitat for amphibians. But as the climate grew hotter and drier, amphibians found they could not adapt. The domination of the reptiles began, culminating in the age of the dinosaurs, which was to last for 165 million years. To put this in perspective, our own species has been around for less than a quarter of a million years.

The ancestors of the

A brachiosaurus, one of the largest known dinosaurs, next to a human for scale dinosaurs were the thecodonts ('socket teeth'), creatures resembling modern crocodiles. They had powerful tails for swimming, and strong hind legs for lunging at their prey. When some of them took to land, they walked on these powerful hind legs, using their large tails as a counterbalance.

The first true dinosaurs (the word is Greek for 'terrible lizard') appeared in the Triassic period, about 230 million years ago. Dinosaurs were one of the most successful and diverse groups of animals ever. There were carnivores and herbivores, swamp- and plains-dwellers, solitary animals and herd animals, small creatures barely larger than a chicken, and huge monsters more than 30 metres long, the height of a four-storey building, and weighing in excess of 100 tonnes.

Other groups of reptiles, the plesiosaurs and ichthyosaurs, dominated the seas, while the sky was the realm of the pterosaurs, flying reptiles with wingspans up to 10 metres across. But other creatures were taking to the air, creatures with feathers. These were the ancestors of the modern birds, derived from a group of warm-blooded dinosaurs. One of the earliest 'missing links' is the fossil of *Archaeopteryx*, a winged creature dating from 154 million years ago. This has reptilian features such as teeth, and birdlike features such as true flight feathers. So, rather than becoming entirely extinct 66 million years ago, some dinosaurs live on in the form of birds.

Warm-blooded and cold-blooded

Many of the dinosaurs, like modern reptiles, fishes and amphibians, were poikilothermic ('cold-blooded'). Cold-blooded creatures need to control their body temperature by external means, for example by warming themselves in the sun before becoming active after a period of dormancy at night. This means they cannot adapt to as wide a range of climatic conditions as homeothermic ('warm-blooded') animals, such as some groups of dinosaurs, together with birds and mammals. Warm-blooded animals can warm up through activity (e.g. shivering), and lose heat via sweating and panting.

MASS EXTINCTIONS

The sudden disappearance of the dinosaurs 66 million years ago is just one of a number of mass extinctions that have occurred in the history of life on Earth. Species become extinct all the time – that is the nature of evolution. But at certain points there have been huge spikes in the rate of extinction.

If we define a mass extinction as the sudden loss of 50 per cent or more of animal species, then there have been five mass extinctions over the past 540 million years. Many other significant extinction events have fallen short of 50 per cent.

These mass extinctions have been identified via the fossils of multicellular organisms, so it is possible that there were earlier ones involving single-celled organisms, which leave little in the way of fossil traces. We do know that around 2.4 billion years ago the accumulation in the atmosphere of oxygen – pumped out by more and more photosynthesizing microorganisms – proved fatal to vast numbers of other microorganisms for whom the gas was poisonous.

The causes of more recent mass extinctions are not always so clear. Any plausible explanation must account for the fact that, although many groups of animals die out, many others survive. Although a catastrophic event may deliver the final blow, this may be preceded by a long build-up of environmental pressures on certain groups of animals. This is the so-called 'press/pulse' model.

Three different types of catastrophe, or 'pulse', are regarded as the likeliest candidates. We know that in the past there have been periods of massive volcanic activity, filling the atmosphere with dust for years on end. This would cut out much sunlight and inhibit photosynthesis, and hence the food supply at the base of nearly all food chains. Volcanic eruption also blasts large amounts of sulphur dioxide and carbon dioxide into the atmosphere. The sulphur dioxide would result in poisonous acid rain, and the carbon dioxide in global warming.

The second candidate is a fall in sea levels, most likely the result of global cooling, as happens during ice ages – when more of the oceans' water is locked up in icecaps. Falling sea levels reduce the area of the continental shelves, the most productive zone of the oceans, and would also have disrupted weather patterns.

The final candidate is the most dramatic. In this scenario, a large asteroid or comet strikes the Earth. The massive explosion would create a hugely destructive shockwave, and possibly also megatsunamis and extensive forest fires. As in the volcanic scenario, the atmosphere would fill with smoke and dust, blocking out sunlight and so disrupting food chains. If the object hit sulphur-rich rocks, this could bring widespread acid rain. It is now widely accepted that a large asteroid did hit the Earth about 66 million years ago, but debate goes on as to whether this was the sole cause of the extinction of the dinosaurs.

Mass extinctions give a spur to evolution, leaving many ecological niches empty. With the extinction of the dinosaurs, a group of small, inconspicuous animals that had been around for over 150 million years seized the opportunity to thrive, diversify and radiate across the planet. These were the mammals.

The final cause?

We may be in the midst of another mass extinction. Some scientists estimate that up to 140,000 species (many of them as yet unrecorded plants and invertebrates) now become extinct every year. The cause? Human activity.

THE COMING OF THE MAMMALS

The first true mammals appeared only 10 million years later than the dinosaurs. But throughout the long reign of the giant reptiles, these early mammals remained similar in size and lifestyle to rats and shrews. Their warm blood – not common among the dinosaurs – enabled these early mammals to function at night, exploiting food sources unavailable to the dinosaurs. It may have been their small size, and consequent need for less food, that enabled them to survive the cataclysm that wiped out the dinosaurs.

The word 'mammal' comes from Latin *mamma*, 'breast', alluding to the fact that infant mammals feed on milk produced in their mother's mammary glands. Another characteristic common to nearly all mammals is the possession of fur. A third common feature is the relatively large size, compared with other vertebrates, of the cerebral cortex, that part of the brain associated with intelligence.

There are three main groups of mammals: monotremes, marsupials and placental mammals. The monotremes are today confined to Australasia and number just the duck-billed platypus and two species of spiny anteater. They do not have external teats; milk oozes out of the opening of the mammary glands, and the infants lap it up from the mother's fur. Monotremes have two other features that are unique in mammals, but shared with their reptilian ancestors: they have a single orifice for excretion and reproduction; and they lay eggs. The eggs hatch after about ten days, and the undeveloped hatchlings spend three or four months being suckled by their mother. The earliest mammals may have been monotremes.

The young of marsupials are live-born, but, like monotreme hatchlings, they are also very undeveloped when they first appear. Marsupial young do much of their development in the mother's pouch, where they have access to her teats. Marsupials today – including kangaroos, koalas, wombats and possums – are confined to Australasia and the Americas. In the past they were much more widespread and diverse, including top predators such as *Thylacosmilus* (a big sabre-toothed cat) and the Tasmanian wolf, which only became extinct in the early years of the 20th century.

Today's most diverse and widespread group of mammals are the placentals, so called because the placenta supplies the foetus with nutrition and oxygen in the mother's womb. The foetus remains in the womb for a long time relative to marsupials, and is therefore much more developed at birth. While in some species (such as humans) the newborn are nonetheless helpless, and their parents must guard and rear them for years, in others (such as antelopes) the newborn can stand straight away, ready to run alongside their mothers.

Placental mammals have adapted to every kind of habitat, from high mountains to dense forests, and from the Arctic to the Tropics. Some, like the whales, have even returned to the oceans, while others, the bats, have mastered the air. But there is just one mammal that has managed to adapt to virtually every climate and habitat. That is because it relies not on fur but on clothing suited to different climatic conditions, and on making fires, and shelters, and using tools rather than claws and teeth to catch and kill its prey.

'Humans arose . . . as a fortuitous and contingent outcome of thousands of linked events, any one of which could have occurred differently and sent history on a different pathway . . .'

Stephen Jay Gould, 'The Evolution of Life on Earth',
Scientific American, (October 1994)

WHERE DO WE COME FROM?

Humans are primates, an order of mammals that also includes lemurs, lorises, monkeys and apes. In fact, humans *are* apes. We share more than 98 per cent of our DNA with chimpanzees and their close relatives, the bonobos.

Primates have highly dexterous hands, and many also have dexterous feet. In most, the thumb can be opposed to the other fingers, enabling them to grasp and manipulate objects – a prerequisite of tool use. Their eyes are large and face forward, giving them good binocular vision – essential for judging distances. Primates' brains are relatively large compared with those of other animals, and this endows them with a great capacity to learn and adapt. The young stay with their mothers for longer than the young of many other animals, and have ample time to pick up skills and customs. Many primates live in complex social groups.

The first primate-like creatures emerged around the time of the extinction of the dinosaurs, 66 million years ago. They resembled squirrels or tree shrews in both size and appearance. The first true primates appeared some 10 million years later. Similar to lemurs and lorises, they spread to many parts of the world, but when the monkeys emerged 34 million years ago they were largely outcompeted. Today lemurs are confined to the island of Madagascar, which the monkeys never reached. The first apes appeared 23 million years ago, but the split between our ancestors and the ancestors of chimps and bonobos did not happen till about 7 million years ago.

By then, large areas of tropical forest had been replaced by more open woodland and savannah. With the change of environment there came a need to adapt to a ground-based rather than tree-based way of life. Around 6 million years ago, early humans began to walk some of the time on their hind legs. This ability, called bipedalism, meant they could look over the long grass to see both predators and prey. It also reduced the surface area of skin exposed to sunlight, and lengthened the stride, so that humans could cover greater distances. The effect was both to widen the range in which they could hunt and gather food, and to enable populations to migrate to different territories altogether.

Our closest relatives

We share 98.7 per cent of our DNA with chimpanzees and bonobos, but these two species differ in their behaviour. Chimps are male-dominated, hunt in groups, are aggressively territorial, and may kill other chimps. Only high-ranking males get to mate. Chimps use a variety of tools, for example to crack nuts or to catch ants. Tool use has only been observed among bonobos in captivity.

Bonobo groups are dominated by females (who have strong bonds with each other), although there is much less sexual differentiation than in chimps. The territories of different bonobo groups overlap, and they have not been observed hunting in groups. Sex is frequent between males and females, and with members of the same sex. Sex is important for social bonding and conflict resolution, not just for reproduction. This has been described as 'sex for peace'.

Claiming that behaviour is genetically determined is always going to prove controversial. But it is certainly possible to see some aspects of human behaviour reflected in that of the chimps, while other aspects are closer to that of the bonobos.

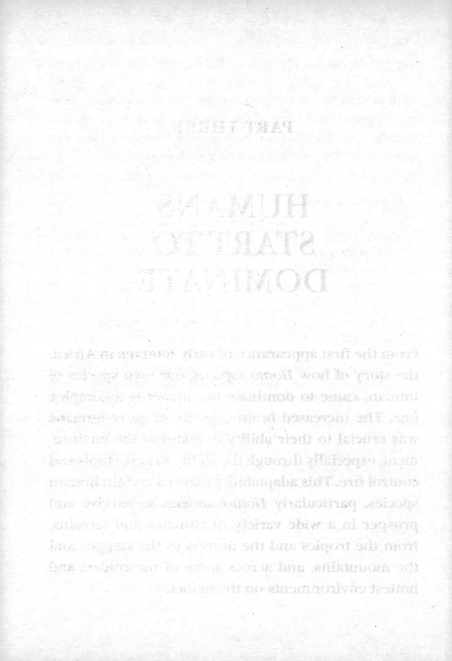

PART THREE

HUMANS
START TO
DOMINATE

From the first appearance of early humans in Africa, the story of how *Homo sapiens*, our own species of human, came to dominate the planet is a complex one. The increased brain capacity of early humans was crucial to their ability to adapt to the environment, especially through the ability to make tools and control fire. This adaptability allowed certain human species, particularly *Homo sapiens*, to survive and prosper in a wide variety of climates and terrains, from the tropics and the deserts to the steppes and the mountains, and across some of the coldest and hottest environments on the planet.

TIMELINE

200,000 years ago: First evidence of *Homo sapiens* in Africa.

150,000–50,000 years ago: Development of language.

100,000 years ago: *Homo sapiens* starts to move out of Africa; early burials with grave goods.

75,000 years ago: Pierced shell necklaces.

45,000 years ago: First fully modern humans in Europe.

42,000 years ago: Flutes of wood and bone appear in Europe.

40,000–35,000 years ago: Human, animal and human–animal figures in Europe, carved in stone and ivory.

38,000–35,000 years ago: Cave art already highly developed.

22,000 years ago: Peak of last ice age.

19,000 years ago: Evidence of wild cereal gathering in the Middle East.

14,000 years ago: Dogs known to have been domesticated from wolves, though this may have happened earlier. First use of grindstones in Middle East.

13,000 years ago: Earliest known portable art in China – engraved antler found in Longyn Cave.

12,000 years ago: Glaciers retreat in Europe.

8,000 years ago: Wheat and barley cultivation spreads to Nile valley from Middle East.

7,000 years ago: Hunting and fishing villages in Yangtze river delta in China start to cultivate rice. Cereal-farming villages established in western Europe.

4,500 years ago: Evidence of long-distance trade throughout South America.

HUMANS PAST AND PRESENT

Fossil finds over recent decades reveal a bewildering range of early human species, many appearing first in Africa. Only some of these were our direct ancestors. The others simply became extinct. We sit at the end of just one of the branches of a tangled family tree.

As early humans spent more and more time on the ground rather than in the trees, bipedalism – walking on two feet rather than four – had by 4 million years ago become the norm. Early humans evolved a number of anatomical adaptations to the new way of walking. For example, the legs grew longer and stronger than the arms, in order to support the full body weight. No longer needed for walking, the hands became better able to hold and manipulate items from food to tools and weapons.

The earliest evidence of tool use comes from 2.6 million years ago, and for the next 2 million years humans used simple stone flakes and cores (and later tools made from bone) for cutting, pounding and crushing. Such tools enabled them to exploit a range of new foods, and to cut meat from larger animals.

The first known member of our own genus, *Homo*, appeared 2.4 million years ago. This was *Homo habilis* ('handy man'), so-called because when its fossils were first discovered in Tanzania's Olduvai Gorge in 1964 this species was thought to be the first tool-user. *Homo habilis* lasted for about a million years, but turned out to be an evolutionary dead-end.

We can't tell for sure when the first humans left Africa, but we know that by 1.6 million years ago another species, *Homo erectus*, had reached South-East Asia as far as Indonesia and China, having first appeared in Africa about 300,000 years earlier. *Homo erectus* was hugely successful, surviving until 143,000 years ago. They were the first species to use fire and cook meat, and there is evidence that they cared for the old and the weak.

About 700,000 years ago one branch of *Homo erectus* began to evolve a larger brain. This was *Homo heidelbergensis*, the first human to make its home in the colder areas of Europe, although some populations remained in Africa. *Homo heidelbergensis* made sophisticated stone flakes and used wooden spears to hunt big game. Its descendants in Europe were the Neanderthals (*Homo neanderthalensis*), while the population that stayed in Africa evolved into modern humans (*Homo sapiens*). Both species appeared about 200,000 years ago, but modern humans did not leave Africa for another 100,000 years.

The Neanderthals: rivals or ancestors?

The Neanderthals were generally shorter and stockier than modern humans, but otherwise they were similar, and their brains were actually larger. They buried their dead, adorned themselves with items like bead necklaces, and were the first humans to wear clothes – essential in the cold climatic conditions then prevailing in Europe. It is also likely that they possessed language.

The Neanderthals disappear from the record between 40,000 and 30,000 years ago. At one time it was thought that their demise began after modern humans moved into Europe, around 45,000 years ago, and that the Neanderthals were either outcompeted or exterminated. But recent studies have shown that 2 per cent of the DNA of many modern human populations outside of Africa is shared with the Neanderthals.

The inevitable conclusion is that over thousands of years the two species interbred, and that many of us must have at least some Neanderthal ancestors.

WHAT MAKES HUMANS HUMAN?

What, if anything, makes humans different from other animals? For centuries, if not millennia, humans have never doubted their own superiority, and insisted that the differences are differences of kind, not just of degree.

Although some human cultures see people as part of nature, the Judaeo-Christian view – that God created humans in his own image and gave them dominion over the Earth – came to dominate. We now know that humans evolved from the same ancestors that gave rise to other living primates, from lemurs to chimpanzees. There is no one point in evolutionary history that we can identify as the moment when humans became different in kind to other animals.

Yet we continue to cling on to this sense of our own exceptionalism. A number of features have been claimed as uniquely human, from consciousness, mind and free will, to language, technology and culture. But science is increasingly showing that none of these assertions are tenable.

Consciousness is our awareness not just of our surroundings, but of ourselves. It is by definition subjective – an internal state known only to its possessor. But scientists have found objective correlations of consciousness, in the form of observed behaviour and brain activity, not just in humans, but also in mammals, birds and even octopuses.

Aspects of consciousness, such as intentional behaviour, making choices and self-recognition, have been widely observed in non-human animals. One simple test uses a mirror to see whether the animal in question knows it is looking at itself, rather than another individual. A number of different primates have passed the test, as well as Asian elephants, bottle-nosed dolphins, orcas and the Eurasian magpie.

Tool use turns out not to be uniquely human either. Chimpanzees poke twigs to 'fish' for ants, sea otters use stones to dislodge and break open shellfish, and a species of crow living in New Caledonia whittles branches into hooks in order to extract food from inaccessible crannies.

'Man is an invention of recent date.'

Michel Foucault, *The Order of Things* (1966)

Whether such behaviours are instinctive or learned cannot always be judged. If they are learned, then we can talk about that species acquiring a culture (see p. 55). A well-known instance of animal culture involved a group of Japanese macaques. One individual started washing sweet potatoes in the sea before eating them, rather than brushing the sand off, as her companions did. Others began to copy her, and this behaviour was passed down the generations.

The vocalizations of various species of whale and dolphin change from group to group, so each 'song' seems to reinforce the identity of that group. The songs also change over time. We don't know if these songs hold sufficient information to count as language – no one has yet established if they have 'meaning'. The same uncertainty applies to other animal vocalizations. Although chimpanzees have been taught to use sign-language, sceptics have pointed out that the failure of any signing chimp to ask a question suggests that this trait is unique to humans. However, over thirty years animal psychologist Irene Pepperberg taught Alex, an African grey parrot, some basic English, and also to distinguish various colours, shapes and sizes. Alex eventually asked what colour he was. Having been told the answer six times, he learned that he was 'grey'. This is the only known example of a non-human asking an existential question. Nevertheless, the line between human and non-human is blurred.

CULTURE

To anthropologists and historians, the term 'culture' encompasses all those components of behaviour that are not instinctive, but consciously created and passed on. So any behaviour that is learned is cultural.

Instinctive behaviours are those that are genetically programmed, and thus common to all members of a species. Newly hatched turtles automatically make their way down the beach to the sea; spiders do not need to be taught how to spin intricate webs. Instincts in humans – shared with other animals – include the drive to eat, sleep, reproduce and nurture offspring.

Memes

The biologist Richard Dawkins coined the term 'meme' for the cultural equivalent of a gene. A meme is any transmittable idea, behaviour, style or technology. Some memes – such as clay writing tablets – flourish for a while until superseded by something better. Others, such as the concept of God, have proved more persistent.

Although most behaviour in other animals is instinctive, not all of it is – for example the ability of chimpanzees, crows and certain other animals to make and use tools. What gives humans an advantage is the sheer complexity of our cultures. The build-up of skills and technologies has allowed humans to adapt to a wider range of habitats than any other animal. In cold climates, they did not grow thick fur coats and layers of blubber, but instead invented and transmitted clothes, shelter, tools, hunting techniques, and so on. Cultural evolution has reduced the impact of natural selection on our species, as weaker individuals are more likely to survive and breed, so slowing the pace of physical evolution.

Culture has given humans an enormous competitive advantage. At the end of the last ice age there were perhaps 10 million humans living on Earth. Today, a mere 10,000 years or so later, the global population is over 7 billion. The rate of human cultural evolution has steadily accelerated, especially in the last 10,000 years, starting with the beginning of agriculture (see p. 80).

Culture and natural selection

Sometimes a cultural innovation can act as a driver of physical evolution. About 7,500 years ago a mutation arose in cattle herders living in central and south-eastern Europe which stopped lactose intolerance in adults. Previously, no humans were able to digest milk and milk products once they were weaned. Now lactose-tolerant individuals could exploit another food source, so drinking milk became a widespread new cultural practice that conferred a competitive advantage. The lactose-tolerant gene successfully spread, and is now found in many populations around the world – but not among cultures that do not raise cattle or other milk-yielding livestock.

Another example is the gene that causes sickle-cell anaemia. Sickle-cell anaemia is a painful disease that causes organ damage. But the same gene provides increased protection against an even more dangerous disease, malaria. That is why sickle-cell anaemia is relatively common in Africa, and especially among yam farmers. To grow yams, the farmers clear forest, and this increases the amount of standing water – ideal conditions for mosquitoes to breed.

With agriculture came new, more complex and more hierarchical systems of social and political organization. Agricultural surpluses allowed some people to live in cities, and to develop specialisms and crafts that were not directly involved in food production. This in turn permitted the rate of technological and intellectual development to speed up – a process that continues to this day. The result has been a huge surge of changes in a blink of evolution's eye, the long-term effects of which are impossible to judge from the heart of the surge. Once every member of a human community – a band of hunter gatherers, say, or a village – could know and touch every other. Today there are cities of tens of millions, nations of billions, enterprises that span the world.

HOW HUMANS POPULATED THE WORLD

Humans have permanently colonized every continent apart from Antarctica. But we did not evolve separately in different parts of the world. All of us alive today, members of the species *Homo sapiens*, can trace our ancestry back to Africa. So when – and how – did we come to spread around the whole world?

There is some uncertainty about when modern humans began to migrate out of Africa, but it was some time between 100,000 and 75,000 years ago. They could spread so effectively because their technologies – tools, clothes, language, disciplined hunting cooperation, use of fire and shelter – were more sophisticated and effective than those of earlier humans, so they were better equipped to adapt to a range of different habitats.

Whether there was one single migration out of Africa or several waves is unknown, but it seems likely that humans moved slowly – perhaps a kilometre or two per year – along the coasts of southern Asia. Fossil remains tell us that by 50,000 years ago they had arrived in Australia. This

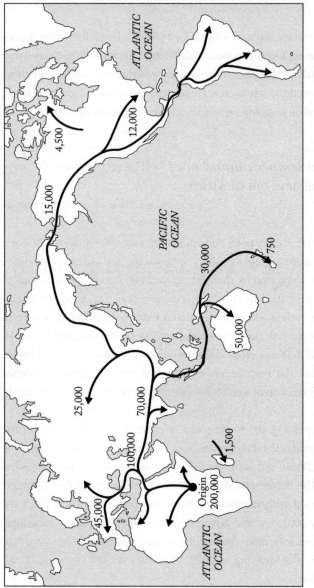

Early migrations of modern humans, with figures showing the number of years ago that they took place

was during the last ice age, when a lot of water was locked up in the ice caps. Falling sea levels exposed a land bridge between New Guinea and Australia, making that part of the journey straightforward. But how these humans crossed the sea to New Guinea remains a subject of speculation – no remains of seagoing vessels survive from such an early date.

'*Ex Africa semper aliquid novi.*' – There's always something new out of Africa.

Pliny the Elder, *Natural History*, VIII (1st century CE)

Although there were earlier human species such as Neanderthals living in Europe, modern humans didn't settle there until about 45,000 years ago – perhaps deterred by the cold climate. The last of the continents to be reached were the Americas, where no earlier human species had ever settled. The earliest evidence of human habitation, found in a cave in Oregon, has recently been carbon-dated to 14,300 years ago. It had been assumed that humans crossed from north-east Siberia via a land bridge situated where the Bering Strait is today, but there is growing evidence that the first colonizers were seaborne, initially settling down the north-west coast.

The last parts of the world to be colonized were the islands of the Pacific. Although the Polynesians had reached Samoa by 800 BCE, the islands of Hawaii and New Zealand were settled less than a thousand years ago. The Polynesians sailed vast distances in twin-hulled outrigger canoes, loaded with families, livestock and plants. In some cases the islands they settled would have been previously visited by fishermen, but others are so remote that the Polynesian seafarers would have had no sure means of knowing whether, when they set sail, they would ever see land again.

THE IMPACT OF THE ICE

About 2.6 million years ago the Earth entered a long period of cooling. There were a number of ice ages, separated by warmer interglacial periods. This period, the Pleistocene epoch, witnessed the evolution of all the known members of our own genus, *Homo*. It may be that the challenging climatic conditions gave a spur to these evolutionary developments, as well as to cultural and technological innovation.

The Pleistocene glaciation is just the most recent of several in the Earth's long history. Scientists do not know what causes them. Irregularities in the Earth's orbit, shifting tectonic plates, alterations in ocean currents and changes in the atmosphere have all been proposed.

During the coldest spells of the Pleistocene, global temperatures fell by 5°C, while the temperature during the interglacials was much like today's. Indeed, we may at present be in an interglacial period, beginning at the end of the last ice age some 12,000 years ago. Some scientists believe a new ice age is long overdue, deferred by human-generated global warming.

During the cold spells, ice caps advanced from the poles and from the high mountains. They covered much of North America, northern Europe and northern Asia, and to the south of the ice there was tundra and permafrost. With so much water held in the icecaps, which were up to 3 kilometres thick, sea levels fell, and left bridges between land masses now separated by sea – such as Siberia and Alaska, or Britain and continental Europe – allowing for the migration of animals, including humans, to new territories. The dry conditions expanded deserts such as the Sahara and the Gobi.

This harsher world drove many animals to extinction. Various mammals adapted by growing thick coats of fur and evolving more massive bodies, the better to retain heat. This 'megafauna' included mammoths, mastodons, cave bears, giant sloths, sabre-toothed tigers and woolly rhinoceroses, as well as still extant species such as wolf, musk ox and reindeer. Some human species, too, grew larger and more robust. Other humans, including our own ancestors, adapted to the more demanding environment by evolving a larger brain, and learning to live by their wits rather than by their strength.

Our cousins the Neanderthals – whose brains were even larger than ours – were well adapted to the permafrost conditions prevalent in ice-age Europe. One cold-weather adaptation was the large Neanderthal nose, which warmed and humidified the cold air they breathed. They made advanced tools and weapons from stone, and must have hunted in teams to bring down large prey such as mammoths. They controlled fire, and cooked both meat and vegetables. Some lived in caves, others in more temporary shelters. One site in Ukraine has turned up dwellings made from mammoth bones and tusks.

The first modern humans reached Europe some 45,000 years ago. They too hunted mammoths, and they gradually replaced the Neanderthals (see p. 52). But even by the end of the last ice age, the whole of Europe probably supported no more than 30,000 humans, while the peopling of the Americas had barely begun.

The extinction of the megafauna

By the end of the last ice age, much of the megafauna – including mammoths and woolly rhinoceroses – had disappeared. A number of explanations have been put forward: overhunting by humans, climate change, disease, and even a comet or asteroid strike. Speculation persists, and it may be that more than one of these factors was involved.

FROM SCAVENGER TO HUNTER

Until they took to agriculture, humans survived on what they could scavenge, hunt and forage in their natural environment. The food resources of any given region tend to be seasonal and limited, so most of our ancestors would have led nomadic lifestyles.

At first, our earliest human ancestors were scavengers rather than hunters. They gathered food from plants and trees, and fed on animals that had died of natural causes or been killed by other predators and the remains discarded. The development of simple stone tools about 2.6 million years ago allowed humans to exploit scavenged carcasses more efficiently. These early tools consisted of a small rock core such as a river pebble, which the maker would strike with another stone to create an edge and sometimes a sharp tip. Using such tools, early humans could dismember the carcass quickly, and take the parts to a safer place for consumption. They could also smash bones to extract the nutritious marrow, and break up tough vegetable matter such as tubers.

The earliest clear evidence for hunting comes from a site in Germany where horses were speared and eaten. This dates from 400,000 years ago, but recent studies of the remains of wildebeest, antelopes and gazelles at a large butchery site in the Olduvai Gorge in Tanzania suggest that early humans (probably *Homo habilis*) may have started hunting very much earlier, perhaps 2 million years ago. It has been conjectured that they might have sat in trees until a herd passed below, and then speared them with sharpened wooden sticks. Some hunters probably relied on running after their prey until the latter became exhausted – a technique still used by some hunter-gatherer groups today.

After *Homo erectus* came on the scene, there was an improvement in stone tools, principally the hand axe. Early *Homo erectus* hand axes were

Killing and cooperation

The ways in which humans have learned to acquire food are closely tied not only to the development of tools, but also to the development of social relations.

In the earlier 20th century anthropologists tended to believe that humans had an instinct to hunt and kill, and that it was this that provided the impetus to develop tools such as spears. They believed that mastery of such tools explained the increase in human brain size.

Today, anthropologists are more likely to argue that it was the advantages of cooperating with each other that led to this increase in brain size. With greater brain size came language and more complex societies. Successful hunting often requires cooperation between individuals, while deciding who is to hunt and who is to gather is an early form of the division of labour.

created using about twenty-five blows, while later ones required around sixty-five blows. *Homo erectus* also used fire to harden the tips of their wooden spears.

Active hunting would have provided a significantly higher proportion of meat in the diet than scavenging. Meat is a dense, protein-rich source of energy, and an increased meat consumption would mean that humans no longer required the long intestinal tracts required to digest raw vegetables and fruit. Food resources could be used more and more to fuel that most important of organs, the brain. And when *Homo erectus* learned to use fire to cook, they could convert food into energy even more efficiently, without having to spend hours chewing it.

FIRE

Fire is the reaction of any combustible material – solid, liquid or gas – with oxygen. It destroys the combustible material, and gives off heat and light. In nature, wild fires are mostly caused by lightning strikes; more rarely by volcanic eruptions.

Various kinds of ecosystem have evolved to cope with regular burning. For example, the seeds of certain trees will not germinate unless they have been subjected to fire, which cracks the hard coating of the seed. The fire will also have cleared the ground of undergrowth, leaving both space and light for the germinated seedlings.

For all animals fire is both terrifying and dangerous. But fire is also essential to human life on Earth. It provides heat, light, and defence against predators. It can also be a means to clear forested areas for agriculture, and to cook food. The ability to tame and use fire was one of the most significant technological breakthroughs made by early humans. When our ancestors first did this is unknowable, but there is evidence that it happened in South Africa as early as 1 million years ago. However, it is not until after 100,000 years ago that we find widespread use of fire by humans.

It takes a lot of work to start a fire. Two flints may be struck against each other to create a spark, or two sticks rubbed together, the friction eventually generating enough heat to set fire to dry grass. More sophisticated techniques involve inserting a stick into a hollow in a flat piece of wood and then using either the hands or a bow to spin the stick very fast. All this takes time and effort, so wandering hunter-gatherers devised various ways of carrying smouldering embers around with them, so fresh fires could be started more easily.

'Fire is a good servant, but a bad master.'

English proverb (early 17th century)

The warmth from fires enabled humans to colonize colder parts of the world. Fire also gave birth to cooking. Perhaps the first cooked meal was produced when a chunk of raw meat fell into the fire. Scientists believe that it was *Homo erectus* who was the first to consume cooked food. This is based on the relatively small size of their molars, compared to those of other apes. This suggests that they may have spent less than two hours a day chewing their food, compared to chimpanzees, which spend a third of their day feeding. Not only could these early humans extract more calories from food if it was cooked, but cooking also enabled them to consume a range of new foods that would otherwise be either inedible or indigestible.

Fire later became a key part of a range of human technologies, from pottery and metal-working to steam power, electricity generation and the internal combustion engine. It has also acquired a range of symbolic significances, as in its use in animal sacrifice and cremation, or as an instrument of eternal punishment, or as an embodiment of purity and truth and love, of passion and inspiration.

HUNTER-GATHERER TECHNOLOGIES

For hundreds of thousands of years, humans relied on simple tools such as hand axes and wooden spears. The next big breakthrough came when they learned to make more complex and effective tools and weapons, enabling them to hunt a wider range of prey species. Finally, having adapted to many different

environments, humans began to reverse the process, adapting their environments to suit themselves.

Thrusting spears made entirely of wood had probably been in use since the start of hunting (see p. 62). These required the hunters to come into close proximity with their prey, which in the case of larger animals could be dangerous. Projectile weapons, those that could be thrown from a distance, lowered the risk to the hunter, and further broadened the range of prey species. A thrown spear also has a higher impact, especially if it has a heavier stone tip attached to its wooden shaft.

Stone points manufactured for spears appear in South Africa about 500,000 years ago, in sites occupied by the common ancestor of *Homo sapiens* and Neanderthals, *Homo heidelbergensis*. These may have been thrusting spears. It had been thought that only *Homo sapiens*, modern humans, were intelligent enough to devise such weapons, but analysis of pointed artefacts made from obsidian (volcanic glass) found in Ethiopia and dating from nearly 280,000 years ago suggests that these were the tips of projectile spears, again the work of *Homo heidelbergensis*.

It was not until the emergence of modern humans in Africa around 200,000 years ago that our ancestors began to adopt a broader subsistence strategy. The great range of tools they left behind suggests that they hunted a wider size range of animals than earlier humans. They also fished. The way such tools were manufactured grew increasingly complex: stone knives made in Europe around 30,000 years ago involved nine different stages and a total of 250 blows to produce. By this time modern humans were making many tools, such as fish hooks and barbed harpoons, out of bone. They also began to make nets, not only for fishing but also to trap small game animals. Bone sewing needles appear upwards of 30,000 years ago. The oldest known bows come from Denmark around 11,000 years ago, but some stone weapon tips of the Magdalenian period, around 20,000 years ago, are so small and light that they may have been made for arrows.

Meat was an important part of the diet, but there was still a reliance on collecting edible vegetable matter such as roots, leaves, nuts and berries, as well as items such as eggs. Modern humans living on coasts also started to exploit shellfish. Contemporary hunter-gatherers allocate more than half of the time spent acquiring food to hunting, over a quarter to foraging, and the rest to processing the food. Until the arrival of agriculture, food processing would have been confined to simpler methods such as grinding, pounding, scraping, roasting and baking. During the last ice age (see p. 60), our ancestors also learned how to store vegetable foods for consumption through the harsh winters.

Modifying the environment

Even before the beginning of agriculture as we know it, humans had begun to modify their environment to increase its food yield. In temperate regions they burned down woodlands to encourage the growth of grasslands, which could support larger herds of prey animals. In tropical regions, people practised 'forest gardening', protecting the most valuable food species and weeding out the inedible ones.

LANGUAGE

Perhaps no other component of human culture, apart from tool use, is as important as language. To be able to communicate complex information is crucial in coordinating all kinds of group activity, from hunting to building a spacecraft. Language is also the principal medium for transmitting other aspects of culture – ideas, technologies, behaviours – via teaching and learning.

Until the birth of writing (see p. 115), language was restricted to speech and signs. Speech of any complexity involves a wide range of different sounds that call for a complex vocal apparatus. Modern humans and Neanderthals seem to have possessed this from the beginning, but fossil evidence suggests that our common ancestors did not.

We do not know how language first arose. Some have suggested that it may have emerged as a more effective way of forging social bonds than the mutual grooming found in other primates. The origins of words may have been imitative, as when a child refers to cows as 'moo-moos'. The word for 'mother' in most languages is similar to 'mama'; the lip movements involved mimic those of a baby seeking its mother's nipple. Group activities would call for conventional noises to indicate what needed to be done – the equivalents of such expressions as 'Hush!' or 'Heave-ho', for example.

'I cannot doubt that language owes its origin to the imitation and modification, aided by signs and gestures, of various natural sounds, the voices of other animals, and man's own instinctive cries.'

Charles Darwin, *The Descent of Man* (1871)

In the 1960s the linguist Noam Chomsky noted how readily young children acquire their mother tongue and suggested that humans have a 'language instinct'. The structural principles of the grammar of any language, he proposed, are universal and hardwired into our genes.

We certainly have the hardware: not only a flexible vocal apparatus, but also a brain equipped for memory and associative learning. But are we, as Chomsky suggests, born with built-in software? The answer appears to be no. 'Feral' children – those reared by animals, or in complete isolation – fail to pick up language. This suggests that children need to have heard a lot of language before they can speak it. And if there really is an innate universal grammar underlying all languages, detailed analysis of our thousands of different languages has failed to detect it. There are all kinds of ways different languages work. Some use a mere eleven different sounds, others up to 144. Rules governing word order vary widely. Some languages dispense with order altogether, instead creating compound words to indicate (for example) who is doing what to whom.

But the multiplicity of languages, and their relations to each other, can tell us much about how modern humans spread around the world. For example, studies of some Siberian and some North American languages point to a common ancestor. Languages form family trees, and these may often mirror genetic family trees – even though language is culturally transmitted. The range of linguistic differences, together with the fact that some languages – such as Basque – bear no resemblance to any other known language, suggests that language may have emerged independently in a number of different places.

KINSHIP

We cannot know for sure how early human groups organized themselves, or how individuals saw themselves in terms of their relations within the group. Anthropologists use the term 'kinship' to denote this web of social relations.

Concepts of kinship vary widely between different societies, indicating that kinship is a cultural construct rather than biologically determined. Although there are, for example, good evolutionary reasons to avoid incest, taboos that govern who may marry whom vary enormously, and often have less to do with genetics than with economics, gender ideas and power dynamics within kinship groups.

The smallest kinship group is the family, but what makes up a family varies widely from culture to culture. In some, it refers to the nuclear family (biological parents and children), in others extended families live together or in close proximity, including grandparents, aunts, uncles, cousins, and so on. In some cultures, marriage is monogamous, in others polygamous – a man takes several wives, or a woman several husbands. Serial polygamy (where individuals take a succession of partners) has become a feature of Western societies, as have same-sex partnerships. Biotechnology has brought new variations – IVF, donor sperm and donor eggs, surrogacy. Humans may also declare those who have no biological connection to be their relations, as happens with adoption. In some cultures, for example among the Inupiat people of Alaska, children can choose who their parents are. In some parts of Malaysia, if you eat rice with a person, they become your kin.

Contemporary hunter-gatherers live in bands of just a few families, totalling no more than a few dozen individuals. The different families are allied via marriage, friendship, common descent and common interest. Bands are egalitarian rather than hierarchical, although certain individuals may have higher status owing to gender or age. It is likely that our own hunter-gatherer ancestors had similar systems of kinship.

As societies grew more complex, a greater variety of kinship systems emerged. The idea of descent, for example, is prone to enormous variation. Some societies trace ancestry through the mother (matrilinear descent), some through the father (patrilinear), some through both, while in others the individual can choose to define themselves either through the paternal or maternal line.

Larger groups in which each individual claims the same common ancestor are known as clans (from Gaelic *clann*, 'progeny'). Sometimes it is not a common ancestor that unites the clan, but a sense of common kinship with a totem, a spirit-being associated with a particular plant or animal. In some parts of the world, members of a clan cannot marry each other.

Clans are sometimes regarded as tribes, or, more often, subgroups of tribes. Tribes were the largest social groups before the development of states, and even today (for example in parts of Africa) they identify themselves as independent or outside of states. What unites them is kinship relations and sometimes a sense of common ancestry. They are typically rooted in a particular place (even if they are nomadic), and often have their own language or dialect.

Status

Although hunter-gatherer bands are notably egalitarian, larger social groupings usually display some degree of social stratification in which certain individuals, families or elites have more status – power, prestige, wealth – than others. Clans and tribes typically have chieftains, often associated with success in hunting or war, but they may also honour other important individuals, such as priests or shamans, or people skilled in a particular craft.

EARLY RELIGION

We will never know what our prehistoric ancestors believed, as they left no written records. But there are many indicators – such as ritual disposal of the dead, grave goods, statuettes and cave paintings – that religious behaviour first emerged tens if not hundreds of thousands of years ago.

It is likely that the increase in the brain size of early humans around half a million years ago gave them the capacity for abstract thought. Imagining something that does not yet exist is essential for developing tools, as is a grasp of cause and effect. These are prerequisites for religious belief, but plainly do not constitute religious belief itself. It is unlikely that religious belief could have appeared in any recognizable form until the emergence of symbolic communication, especially complex speech – and this only appeared with the emergence of Neanderthals and modern humans.

There are indications that Neanderthals disposed of their dead in a ritual fashion. Neanderthal remains dating from 65,000 to 35,000 years ago found in the Shanidar caves in Iraqi Kurdistan include the body of a man who may have been buried on a bed of flowers. Elsewhere corpses appear to have been coloured with red ochre. Clearer evidence emerges with the appearance of modern humans, especially in the Middle and Upper Palaeolithic (see p. 90). Bodies were sometimes buried with grave goods, indicating a belief in some kind of afterlife where the dead might use them. Evidence for cremation has been found by Lake Mungo in Australia, dating from 40,000 years ago. One of the earliest known pieces of figurative art, dating from the same period, comes from the Hohlenstein-Stadel cave in Germany. Carved from mammoth ivory, the standing figure has a humanlike body but the head of a lion, and may represent a deity.

From 30,000 years ago, 'Venus' figurines begin to appear at Upper Palaeolithic sites across Europe and Siberia – small sculptures of naked women, only a few centimetres in height, carved out of materials including stone, bone and ivory. The figures have highly stylized legs and heads, and the arms and feet are missing, but the breasts, buttocks, stomachs and pubic areas are depicted in detail, and on an exaggerated scale. This has given rise to the theory that they are associated with some kind of fertility cult, or that they are depictions of a mother or creator goddess.

'Religion is the dream of the human mind.'

Ludwig Andreas Feuerbach, *The Essence of Christianity* (1841)

At least some of the magnificent rock paintings dating from 30,000 to 10,000 years ago found in caves in France and Spain may have had religious significance. Those at Lascaux include mysterious beasts, some half human and half lion, while others are human–bird hybrids. This suggests that the people who painted them may have practised some form of shamanism. In shamanism – still found today among various hunter-gatherer groups – particular individuals take on the role of a kind of priest. The shaman enters a trance, sometimes induced by drumming, sometimes by taking psychotropic plant extracts. In this trance the shaman may take on the identity of an animal and go on a spirit-journey to ensure good hunting. Shamans may also claim other supernatural powers, such as divination and healing.

THE BEGINNING OF ART

Art – in the sense of creating objects whose value is purely aesthetic – is a fairly recent cultural construct. In the past people would have thought of 'art' not as an end in itself, but as the craft involved in producing objects with a social function.

That function may have been to do with ritual, for example, or religion, or group identity, or personal status. This still holds true in many societies round the world. Today in the West, for example, art objects are frequently collected by the wealthy as status symbols or financial investments.

When 'art' began depends on what we mean by the word. Early humans were making objects such as simple hand axes more than 2 million years ago (see p. 62), but these objects appear to have had a purely practical purpose, and had no decorative element, no indication that they referred to anything other than themselves – they were simply tools.

The Lion Man of Hohlenstein-Stadel

Conversely, some particularly fine and fastidiously polished Neolithic stone axes weren't made for clearing forest for farming, but to be buried among the owner's grave goods as a marker of status.

It was long thought that the earliest 'art' was created by modern humans in the Middle to Upper Palaeolithic, starting about 40,000 years ago. Objects such as the Lion Man of Hohlenstein-Stadel, the 'Venus' figurines, and the Lascaux cave paintings of animals are thought likely to have served some kind of religious purpose (see p. 73). Animal paintings from the same period have been found in Sulawesi, Indonesia. All these items clearly depict or refer to recognizable objects. But recent discoveries show that non-representational 'art' may be much older.

In 2013 archaeologists found a gridlike design carved into the wall of Gorham's Cave in Gibraltar. The design, a sort of hashtag, dates from at least 39,000 years ago. What was remarkable was that it had been made by Neanderthals. Researchers established that the painstaking work could not have been the by-product of butchering animals. The only alternative, they believe, is that it was symbolic, indicating that Neanderthals had a capacity for abstract thinking.

Just a year after the discovery in Gorham's Cave, the world of palaeo-anthropology was shaken again when researchers examined a mussel shell found at a site in Java and detected a faint zigzag pattern. What was astonishing was that this pattern had been carved by *Homo erectus* 500,000 years ago. Again, experiments showed that creating such a 'doodle' was both difficult and deliberate. Whatever it meant to its creator, this is an object that revolutionizes our understanding of when our distant ancestors began to be capable of abstract thought.

'Art is not a mirror to reflect the world, but a hammer with which to shape it.'

Vladimir Mayakovsky, Soviet poet (attributed remark)

SHELTER

Many species build shelters. Some animal shelters are rudimentary, such as the piles of twigs that pigeons nest in. Others are extraordinary feats of engineering, from the burrows and dams built by beavers to the massive towers raised by termites, with their mazes of passages, brood chambers and gardens, and remarkable ventilation systems.

The first shelters built by our remote ancestors would have been extremely crude in comparison. They may have resembled the nests that other apes, such as chimpanzees and gorillas, build habitually in trees. But when early humans left the tropical forests and savannahs of Africa and began to spread to cooler climes, their ability to find and / or build more weather-resistant shelters became vital.

Evidence of early human habitation has been best preserved in caves, where humans camped, but did not build formal living structures. But there were too few caves to house all those who lived during the Palaeolithic period, so it is likely that many open-air shelters were built, whose remains have been lost.

The first towns

By 8000 BCE, the settlement at Jericho in Palestine had grown into a small town of some seventy dwellings, housing a few hundred people. The houses were round, and built of bricks. This new building material was made by mixing straw into wet clay, shaping the material into blocks with curved edges, and leaving them to dry in the sun. Once a wall of bricks was built, it was plastered with mud. Most of the houses consisted of a single chamber, but some had up to three. The settlement was surrounded by a wall, perhaps a defence against flooding, and within the wall was a tower over 3.5 metres high and with an internal staircase. This is the earliest example of such a structure, which may have had a ceremonial purpose. At Çatal Höyük in southern Turkey there is a much larger town, which was well established by 7000 BCE, and may have been home to several thousand. The ground plan of the brick-built houses is not round, but rectangular. No ground-level paths ran between the houses, which were accessed down ladders from their flat roofs. The interior walls were smoothly plastered, and smoke from hearths and ovens escaped through the ceiling access.

One of the earliest examples of shelters constructed in the open comes from the Terra Amata site near Nice in the south of France. Here postholes were found, providing evidence of oval wooden frameworks, some nearly 15 metres long and 6 metres wide. These dwellings also had fireplaces. The dating is debatable: estimates range from 380,000 to 230,000 years ago.

Most such huts or tents would have had wooden frames. The materials that covered them varied, from animal hides to reeds daubed with mud to brushwood. The standard ground plan was circular. As the people who built them were nomadic hunter-gatherers, such shelters would have been temporary. It was only with the coming of agriculture (see p. 80) that people began to live in permanent settlements, and to build more durable shelters – houses.

CLOTHING

As well as tools, fire and shelter, clothing was crucial in enabling humans to populate the cooler areas of the world. But clothing materials are perishable, and archaeological evidence is rare. Some of the best-preserved examples come from arid areas, or from the acid peat bogs of northern Europe.

There is indirect evidence that the Neanderthals, who appeared 200,000 years ago, made clothes. It was during the era of the Neanderthals that DNA analysis suggests that body lice (which live in clothing) diverged from head lice (which live on the scalp). Stone scrapers dating from about 100,000 years ago suggest that the Neanderthals probably used these implements to clean the meat off hides. The Neanderthals were certainly intelligent tool users, and hunted big game animals such as mammoths, deer and musk ox. They may have cut the hides into wearable shapes, and made holes for head and arms. It is unlikely they could have survived in Europe through a series of ice ages without developing such technologies.

Modern humans appear to have devised a much wider range of clothing technologies. They too needed to adapt to the conditions of the last ice age in the northern part of their range. Some dyed flax fibres found in a cave in Georgia have been dated to 38,000 years ago, and pieces of bone and ivory found in Russia from 32,000 years ago may have been used as needles. For punching holes in animal skins, sharp awls were used, and these holes were then laced.

Animal hide remained the principal clothing material until the advent of weaving. The discovery in 2008 of fragments of clay with impressions of wool fibres indicates that weaving may have begun as long as 27,000 years ago.

Iceman fashion

In 1991 two walkers crossing a 3,200-metre pass over the Ötztal Alps between Italy and Austria found a corpse half frozen in the ice. Examination by scientists revealed that the semi-mummified body, dubbed Ötzi, dates from around 3300 BCE. It was not only his body that was remarkably well preserved – so were his clothes. These helped to confirm our understanding of how clothing had been made and worn for many thousands of years before his time.

On his lower legs Ötzi wore carefully sewn leggings, while his groin and buttocks were covered in a thin leather loincloth. To keep him warm at this altitude, he wore a fur cap and a long-sleeved, thigh-length coat made from many pieces of fur. On his feet he wore short boots made from animal hide, which he had stuffed with grass to keep out the cold of the snow. But it wasn't the cold that killed Ötzi. It was the arrow wound found in his shoulder, and the blow to the head.

POTTERY

To keep or to carry many foodstuffs – berries, grains, and so on – requires some kind of container. Before the invention of pottery, people used skins, or baskets made of leaves or twigs, but these let vermin in, and could be damaged by fire and water.

The pot, made out of widely available clay, provided a durable solution. Pots are not only relatively sturdy, but you can also cover them, so keeping the freshness in and the vermin out. Pots are waterproof, especially if fired, so will hold liquids. They are also fireproof, so they can be used for cooking food.

It used to be thought that pottery developed alongside agriculture around 10,000 years ago, but fragments found in a cave in southern China turned out to be twice as old, and have what appear to be scorch marks – indicating that the pots were used for cooking by nomadic hunters, long before settled farming. It is quite likely that pottery was invented independently – and also forgotten – in a number of different places at a number of different times.

Early pots usually had round bottoms, as edges are liable to crack. They were made by pinching or by coiling long ropes of clay, a laborious procedure simplified by the invention of the potter's wheel in Mesopotamia around 6,500 years ago.

Firing pottery alters its chemistry and structure permanently, making it much more durable, and resistant to higher temperatures. The earliest known example of a fired pot comes from the Jamon culture of Japan, dating from 7,000 years ago. This might have been accidentally exposed to fire, but subsequently pots were fired in pits, at temperatures up to 900°C.

Pottery, food and drink

Until the advent of the pot, cooking meant either roasting items over the fire, or baking them in the embers. The fireproof, waterproof pot meant that people could now cook by boiling and stewing – techniques that extract nutrition from tough bits of animal carcass that would once have been discarded. The first known evidence of a dish cooked in this fashion dates from 8,000 years ago. It was a soup made from hippopotamus bones. One thousand years later people in Iran were fermenting grains in pottery jars to produce one of the first known beers.

Not all pottery is for practical purposes, and not all pottery consists of pots. Perhaps the earliest examples of non-functional earthenware are the figures of animals found in Croatia, made between 17,500 and 15,000 years ago. Several millennia later, the earliest civilizations – in Mesopotamia, China and India – made decorative tiles, statues and jewellery out of earthenware. These were often brilliantly coloured, the result of combining clays with other minerals and then firing the mixture – glazing. This method required the development of kilns that could reach much higher temperatures than could be achieved by pit firing.

THE FIRST FARMERS

When humans began to farm, they set in motion a revolution in the way that we live. Before farming, humans were nomadic hunter-gatherers. With the advent of farming, people began to live permanently in one place – as the vast majority do to this day.

With farming, not only could people build up stores of food for the winter and for other lean times, they could also trade food surpluses. Swapping these extra resources for raw materials, manufactured goods and labour underpinned a whole range of hitherto undreamt-of activities, from building temples to waging wars. Food surpluses also made it possible for part of the population to live in cities, and for some of these people to practise more highly specialized crafts and activities.

In the tropics, some groups had been practising 'forest gardening' for millennia (see p. 67), and elsewhere hunter-gatherers had begun to collect and store non-perishable foodstuffs such as cereal grains. The next step was to realize that if you sowed these grains in the ground, you could harvest many more than you had planted.

This seems first to have happened between 10,000 and 8000 BCE in the warmer climatic conditions that followed the last ice age. At this time in the 'Fertile Crescent' of the Near East, in the valleys of the Tigris and Euphrates rivers, people began to grow wheat and barley, and later rye and beans – the start of arable farming. Over the next several thousand years, arable farming arose independently in various other parts of the world. Maize, gourds, peppers and potatoes were cultivated in the warmer regions of the Americas, millet in northern and central China, rice in southern China and South-East Asia, and a variety of cereals and root crops such as yams in sub-Saharan Africa.

'People manured the fields and planted cereals. They seized wild animals and made them into domesticated livestock.'

Book of the Master from South of the Huai River,
a Chinese compilation from the 2nd century BCE

The very process of cultivation can produce varieties with better yields. Wild wheat drops its grains once they are ripe, but a random mutation produced individuals whose ripe grains stayed on the plant, making them easier to harvest. Early farmers would have selectively harvested and sown grains from such individuals, so inadvertently helping to propagate a new variety.

A range of new tools and technologies made farming less laborious and increased yields. By 6000 BCE basic crop rotation, alternating cereal crops with beans, was being practised in the Fertile Crescent. Livestock farming also began here (see p. 82), and farmers used rotted-down animal droppings – manure – to make the soil productive. In drier areas, they dug irrigation ditches.

Digging sticks, used to till (loosen) the soil before planting or sowing, and to get rid of weeds, gave way to mattocks and hoes. But even more important was the plough – at its simplest a blade attached to a long handle. Ploughing brings fresh nutrients to the surface, and buries weeds and what's left of the previous crop, which then rot down and enrich the soil. Early ploughs were made of wood, and pulled by hand. Later, draught animals such as oxen, buffalo and horses were coupled to ploughs, and the blades began to be made of a stronger, harder new material – iron (see p. 94).

The spread of agriculture also affected how humans related to the land around them. Land and field boundaries were laid out as settlement became more permanent. And the concept of human 'ownership' of the land (as opposed to territorial claims on hunting grounds) became more common among cultures who adopted a farming lifestyle.

DOMESTICATING ANIMALS

In some parts of the world humans had been practising game management for millennia before they began to domesticate animals for food. For example, in some temperate areas forest was burned to clear the way for grassland, on which prey species grazed.

We will never know exactly how humans first managed to domesticate wild animals, but we can guess that they chose species that were not too aggressive and which had a herd instinct, making them easier to manage. Between 9000 and 8000 BCE, sheep, goats, pigs and cattle were all being farmed in various parts of the world, from China through southern Asia, the Middle East and North Africa. Other animals followed: the guinea pig and llama in South America, the donkey in Egypt and the Near East, the horse on the Eurasian steppes, and the chicken and water buffalo in India and China.

These animals yielded a variety of products, particularly meat, wool and hides. Some, such as donkeys, oxen and llamas, provided muscle power. They carried loads, drew ploughs, and pulled sledges and carts. Some animals, rather than being slaughtered, were bled regularly – blood is a nutritious food. A similar form of animal harvesting emerged after a mutation arose in some human populations that removed adults' lactose-intolerance (see p. 56). Among these populations, cattle, sheep and goats were valued as much for their milk, and the products that could be made out of their milk, as for their meat.

'Man's best friend'

Although in a number of cultures wolves are the epitome of the 'savage beast' they were the first animals that humans domesticated. Tamer, less aggressive individuals found they could feed off food scraps near camp fires. Humans in turn found these tamer wolves could warn them of dangers and help them to hunt, and so was born a mutually beneficial partnership. There were probably many false starts, going back up to 40,000 years ago, but the latest DNA studies indicate that all dogs today descend from a single domestication that occurred between 11,000 and 16,000 years ago, while humans were all still hunter-gatherers. Once the tamer individuals had been adopted, humans would have used selective breeding to bring out characteristics that they prized – hence the wide range of breed types found today.

Neither fresh milk nor fresh meat will keep long without spoiling, but various techniques emerged for preserving such foods. If milk is made into cheese it keeps much longer, while preserving the nutritious fats and proteins of the raw material. It isn't known when cheese was first made, but strainers with traces of milk fats have been found in Poland dating from 5500 BCE. Similarly, meat and fish can be preserved by various methods of curing, including air-drying, smoking and salting. So important was salt that it was traded long distances across Europe, the Mediterranean, Africa and Asia.

Although growing crops and raising livestock helped to keep food in regular supply, the populations who took up farming found themselves mostly relying on a single staple carbohydrate crop, such as maize, rice or wheat. Comparing the skeletons of these early farmers with those of their hunter-gatherer ancestors, it is clear that the latter, whose diets were more varied and richer in protein, were healthier and sturdier. Agriculture supported larger populations, but they were not necessarily healthier ones.

PUTTING ANIMALS TO WORK

Many of the first domesticated animals were reared for products such as meat, wool, hides, milk. But over time humans began to value certain animals for their strength, and for the first time the exploitation of the environment did not rely solely on human muscle power.

Oxen (male cattle castrated to make them more docile) were first harnessed and put to work from about 4000 BCE, in both Europe and the Middle East. At first they dragged sledges, and later ploughs and wheeled wagons, enabling greater areas of land to be cultivated. Their relatives, water buffalo, were similarly used in southern and South-East Asia, being particularly adapted to the wet environment of rice paddies.

The horse – which was to become the dominant mode of transport in many parts of the world for the best part of five millennia – was first domesticated in around 3000 BCE, in the steppes around the Black Sea and the Caspian Sea. Wild horses are relatively small, but selective breeding by humans produced a great variety of different builds and sizes, suited to a wide range of roles, from pulling heavy carts to carrying messages long distances – it was not until the invention of the steam locomotive that the speed of the horse was to be surpassed.

The wild ass was domesticated around the same time as the horse, in Egypt, and both horses and donkeys were widely used by the early civilizations of Mesopotamia and Egypt.

The horse in war

Horses were not at first used in war as cavalry – in which each horse carries an armed rider – but to pull chariots. Typically, these were lightweight carts drawn by one or two horses, and carrying a driver and a single warrior, armed with throwing spears or a bow and arrows. In the 2nd and 1st millennia BCE such war chariots were widely used across Europe, the Middle East and central and southern Asia as far east as China, but by the start of the Common Era had largely been superseded by cavalry. Armed horsemen are more manoeuvrable. They can mass in larger units than chariots, and on rougher ground. At first, cavalry were relatively light, as the horses were too small to mount heavily armoured riders. Riders were typically armed with javelins or bows and arrows. With the breeding of heavier horses, the introduction of the stirrup and more stable saddles, and with heavily armoured riders deploying heavy lances against the enemy's line, cavalry could be used as a shock weapon. Later, the development of firearms eroded the tactical advantage of heavy cavalry.

The third major group of working animals comprises the camelids – the camel family. In South America, the chief working camelid is the llama, which has long ceased to exist in the wild. Although the llama was widely exploited by successive civilizations in South America as a pack animal, it lacks the strength to pull either a plough or a wheeled cart. In the Old World the single-humped Arabian camel (found from North Africa through to India) and the two-humped Bactrian camel of Central Asia and Mongolia are much larger, and are used as draught animals as well as for riding. Camels are well adapted to desert conditions, and can survive long spells without water (although they can drink 100 litres of water in a matter of minutes). The fat in their humps stores energy in times of food shortage. Camels were first domesticated in Arabia, and by 1000 BCE caravans of camels were carrying precious goods up the west coast of Arabia, providing a trade link between India in the east, and Mesopotamia and the Mediterranean in the west.

The last major group of working animals are the elephants. The Indian elephant (once found as far west as Syria) was being used as a beast of burden in the valley of the River Indus by 3500 BCE, and has been used in agriculture and forestry ever since. It was also used in warfare, to carry armed troops. African elephants are far less docile than their Indian cousins. Although they were famously used in battle by the Carthaginian general Hanibal against the Romans in the 3rd century BCE, they often inflicted as much damage on their own side as on the enemy. Attempts to domesticate them were abandoned long ago.

THE WHEEL

The wheel has proved to be one of the most enduring and useful pieces of human technology. It was probably first invented in the 4th millennium BCE, and became a useful means of transporting heavy loads. It remains in widespread and growing use.

For brevity's sake we refer to 'the invention of the wheel', but it is the combination of wheel and axle that comprises the key technology. Of course, the most obvious use is in transport, but numerous other machines rely on wheels, in the form of flywheels, cogged gear wheels, and so on. The first archaeological evidence of a wheel, found in Mesopotamia, and dated to around 3500 BCE, was probably a potter's wheel, although those may have been invented a thousand years earlier (see p. 79).

A wheel is more than just a rolling cylinder. Such rollers, shaped from

Late Iron Age chariot burial, found in Marne, France

tree trunks, were probably in use long before the wheel–axle concept came along, and served to move heavy loads, like large stones, short distances. The wheel–axle combination reduced the friction between the ground and the rolling cylinder, and joined it to a stable platform. The technology brought challenges, however. Each end of the axle, together with the holes in the centre of the wheels, had to be smooth and round, otherwise there would be too much friction for the wheels to rotate. The first wheels were solid slices of wood, before the introduction of the wooden spoke around 2000 BCE made for a much lighter, better-sprung device.

There is some debate as to where and when the first wheel–axle combination was invented. Mesopotamia and various parts of the Eurasian steppes have both been proposed, with dates between 3300 and 3000 BCE. In these parts of the world wheeled vehicles would at first have been pulled by oxen alone, but the domestication of the horse and wild ass around 3000 BCE provided faster pulling power. It was not long before the agricultural cart was adapted for warfare as the chariot (see p. 85).

> 'When man wanted to imitate walking he created the wheel, which doesn't look like a leg.'

Guillaume Apollinaire, French poet, introduction to
Les Mamelles de Tirésias (1917)

Wheeled vehicles came into use in China in the 2nd millennium BCE, and subsequently across much of Eurasia, but were never used by pre-Columbian Americans, probably because suitable draught animals were not available. The llama is not strong enough to pull a cart, while the North American bison resisted domestication (wild horses had become extinct in the Americas around 12,000 years ago).

NOMADS

Our hunter-gatherer ancestors led wandering lives. Once they had exhausted one area of game and other food resources, they would move on. The coming of agriculture encouraged people to stay in the same place, and this led to permanent settlements.

That, at least, is the broad picture. But after the end of the last ice age, as tundra gave way to grasslands and forests, some areas proved so productive that hunter-gatherer groups could stay in place from season to season, from year to year.

Even with the advent of agriculture, certain groups of pastoralists – herders of animals such as cattle, sheep, goats, camels or reindeer – found that in the harsher areas where they lived they needed to keep constantly on the move in search of fresh pasture. This still happens in various parts of the world, although nomads have grown fewer and fewer.

In some areas, movement is not continual but is dictated by the seasons (wet/dry, or summer/winter), which the herders spend in different places. In mountainous areas, this movement may be between two different altitudes, with a permanent home in the valley and a summer home in the high pastures.

A medieval description of nomads

In *The Travels of Marco Polo* (1298) there is a detailed account of the nomadic lifestyle of the Mongols:

> They never remain fixed, but as the winter approaches remove to the plains of a warmer region, in order to find sufficient pasture for their cattle; and in summer they frequent cold situations in the mountains, where there is water and verdure, and their cattle are free from the annoyance of horse-flies and other biting insects . . . Their huts or tents are formed of rods covered with felt, and being exactly round, and nicely put together, they can gather them into one bundle, and make them up as packages, which they carry along with them in their migrations . . . They subsist entirely upon flesh and milk.

Not all nomads are pastoralists. Some are traders, such as Roma horse-dealers or the Tuareg who take their caravans across the Sahara, while others are itinerant craftsmen, such as the Irish Travellers who traditionally mended pots and pans.

Nomads have often found themselves in conflict with sedentary populations. The latter have developed strong ideas of territory and land ownership, whereas nomads take a very different view, and tend to ignore boundaries and borders. Today the balance of power is very much in favour of the sedentary populations, who frequently resent and discriminate against nomads. Modern states, believing that nomads are ungovernable, often apply pressure on them with the aim of getting them to settle in one place.

In the past, the balance of power was sometimes very different. The vast steppes that stretch across much of eastern Europe and central Asia once cradled a succession of warlike nomadic horsemen – Huns, Magyars, Mongols and others – who spread in all directions, from China in the east to Hungary and even further west, looting and destroying cities and slaughtering their populations (see p. 144).

FROM STONE TO BRONZE

For more than 2 million years, humans relied on readily available raw materials to make their tools – wood, bone and, above all, stone. Such was the importance of stone that archaeologists characterize this period as the Stone Age.

The Stone Age is divided into periods. The Palaeolithic (Old Stone Age), ending around 10,000 BCE, was followed by the Mesolithic (Middle Stone Age), and then the Neolithic (New Stone Age). It was the Neolithic that saw the birth of agriculture (see p. 80), a fundamental revolution in the way that humans sustain themselves. It started about 9000 BCE in the Near East, and independently around the same time in China. Over the next 3,000 years the Neolithic had appeared in North Africa, central Europe and southern Asia. It did not reach western Europe until 4000 BCE.

The Neolithic brought Stone Age technology to its peak. As in the Palaeolithic, some tools were still made by flaking materials such as flint and obsidian (volcanic glass), but using many more stages and blows to produce items such as knife blades. Other tools, such as the stone axes used to clear forests for farming, were made by polishing or grinding

The Porsmose Man, a Neolithic skull from Denmark with an arrowhead still embedded in it

coarser-grained rocks such as basalt, jade and greenstone against an abrasive stone, using water as a lubricant. Such flaking and polishing techniques were labour-intensive and were practised by specialized artisans whose products were highly valued and widely traded. Although a few objects were made of copper during the Neolithic, metal did not come into widespread use for tool-making until the Bronze Age, which arose independently around 3300 BCE in the Near East, 3200 BCE in South-East Europe, and 3000 BCE in China. The technology spread westward and northward across Europe, reaching Britain around 2000 BCE.

Bronze is an alloy, a mixture of copper and a small amount of another metal, typically tin. Copper is only rarely found in its elemental form, and must usually be extracted from its mineral ore by smelting, which calls for high temperatures. The alloying process also involves heating and melting the metals.

Copper on its own is quite a soft metal, but the addition of tin makes it much harder – harder even than the stone tools that preceded it. Bronze made possible not only the manufacture of harder and sharper axes, but also a range of new tools and weapons, from swords to breast-plates and helmets.

Tin is rarer than copper, and the two metals are not often found close together, so Bronze Age technology promoted long-distance trade. For example, tin from the mines of Cornwall was exported as far as Phoenicia in the eastern Mediterranean. Another rare metal, gold, also came into use, largely for decorative purposes.

'When the age of decadence arrived, people cut rock from the mountains, hacking out metals, smelting copper and iron ores.'

Book of the Master from South of the Huai River,
a Chinese compilation from the 2nd century BCE

Perhaps because objects of bronze and gold were rare and precious, Bronze Age cultures showed sharper social distinctions than had pre-vailed in the Neolithic. The bronze sword in particular brought the for-mation of elite warrior castes.

Status and power were also embodied in grand building projects, from the Step Pyramid of Djoser in Egypt, of the 27th century BCE, the earliest building of its size in the world, to the palaces of the Minoan and Mycenaean cultures in Crete and mainland Greece, and from complex

The Ring of Brodgar, Orkney

religious sites such as Stonehenge in England, to the magnificent royal tombs of the Shang dynasty in China. Stone alignments and circles were found in many cultures. In the British Isles, where they date from about 3200 to 1500 BCE, they became progressively more complex, suggesting a tendency towards ritual and perhaps political centralization. The religious beliefs that inpired their creation are obscure, though advanced astronomical knowledge is in evidence: the midsummer Sun rises along the axis of Stonehenge. Similar ritual centres elsewhere such as the vast and complex tombs of the Boyne Valley in Ireland or the structures around Maes Howe and the Ring of Brodgar in Orkney, would have each required hundreds of thousands of man-hours to construct. They demonstrate large-scale command activity, and therefore greater social stratification.

FROM BRONZE TO IRON

Like the Bronze Age, the succeeding Iron Age – when iron became the principal material for making tools and weapons – began at different times in different regions. The earliest evidence of iron working comes from the Near East, and dates around 1200 BCE.

Iron working appears to have developed independently on the Indian subcontinent around the same time, and a little later in China. It spread into Europe from the Near East, probably via the Caucasus, and had spread across the entire continent by 500 BCE. Some places, such as sub-Saharan Africa, skipped the Bronze Age altogether, as iron directly replaced stone.

Before the Iron Age, the only iron in use was that found in meteors in its elemental form, but this was rare and used only for decorative objects such as beads. It was not until people learned to extract elemental iron from its mineral ore by smelting that a technological revolution became possible.

Wrought iron is generally softer than bronze, so iron tools wear out faster. But iron took over from bronze partly because sources of iron ore are much more widespread than sources of copper and tin, and partly because iron implements are cheaper to produce. Iron hoes and iron nails were important innovations in agriculture and construction respectively.

The technique of adding carbon to iron to make steel – which is harder and stronger – was certainly known by the time of the Romans. The proportion of carbon was critical: add too little, and the iron is not hard enough; too much, and it turns brittle. This made steel items more expensive, and wrought iron stayed in use for cheaper items.

The Bronze Age created societies with small elites because the technology that gave people power was so expensive. In the Iron Age, with iron tools and weapons more widely available, power was more evenly distributed, although there was still some social stratification. The difference can be seen in modes of warfare in ancient Greece. Homer's *Iliad* describes the Trojan War (set during the Mycenaean Bronze Age) in terms of single combat between kings and princes, who ride into battle in chariots. But following the rise of the Greek city-states around 750 BCE, wars were fought by all adult male citizens, who had to supply their own arms and armour, and who fought as infantrymen in well organized formations.

The Celts

In several parts of the world – the Mediterranean, and across south-western, southern and eastern Asia – iron-working technology was developed by societies that were already partly urbanized. However, societies in temperate Europe during the Iron Age were still pre-urban and largely tribal. Many of these, especially in western Europe, have come to be known as 'the Celts'. But although they had linguistic and cultural elements in common, they would not have identified themselves as Celts, and DNA analysis shows that there was great genetic diversity. These tribes were village-based peasant farmers. Villages may have had headmen, but kinship is likely to have played an important role in social organization. Iron tools enabled them to clear forest faster than before, and they also drained marshes. The numerous hill forts typical of the period indicate that war was not uncommon, but these forts were probably only occupied in times of danger. When the Romans invaded Gaul (France) and then Britain, the hill forts proved inadequate defences against the organized military might of an established urban civilization.

CIVILIZ

PART FOUR

CIVILIZATION

In certain places, agriculture produced tradable food surpluses. These helped to underwrite a wider range of human activities than had been seen hitherto, from increasing craft specialization to organized religion. Surpluses also helped to fund the building of cities, and the creation – often through warfare – of states and empires. As human societies became more complex, writing and law became essential instruments of administration, while increasingly elaborate hierarchies of power were maintained through force.

TIMELINE

5500–4000 BCE: Sumerian civilization established in the Euphrates valley in southern Mesopotamia (now Iraq).

3650–1400 BCE: Minoan and other early Aegean civilizations.

3100 BCE: First pharaoh unites Upper and Lower Egypt.

2600–1900 BCE: Mature period of the Indus valley civilization.

2580–2560 BCE: Great Pyramid of Giza built.

2070 BCE: Xia dynasty in China.

2000 BCE: Preclassic period of Mayan civilization in Mesoamerica.

1754 BCE: Code of Hammurabi, an early legal system, created in the Babylonian empire.

1650 BCE: Hittite kingdom emerges in Turkey.

1600 BCE: Shang dynasty emerges in the middle valley of the Yellow River in China.

1600–1500 BCE: Olmec civilization established in modern-day Mexico.

1550–1077 BCE: New kingdom of Egypt rules empire stretching from the Levant to Nubia.

1500–800 BCE: Vedic Age: creation of ancient Hindu scriptures in India.

1070 BCE: Kingdom of Kush established in modern-day Sudan.

1000 BCE: Golden age of Phoenician cities, including Tyre and Sidon.

911–612 BCE: Neo-Assyrian empire in the Tigris valley.

900–200 BCE: Chavin civilization in modern Peru.

800–400 BCE: D'mt kingdom in Ethiopia.

***c.* 550 BCE:** Cyrus the Great of Persia founds Achaemenid empire.

510–323 BCE: Classical period of ancient Greece.

509 BCE: Roman republic established.

331 BCE: Alexander the Great of Macedon defeats Persia's Achaemenid empire and goes on to rule from the Adriatic Sea to the Indus River, spreading Hellenistic influence far and wide.

321–185 BCE: Mauryan empire in India.

300 BCE: Construction of the library of Alexandria, the ancient world's largest.

221 BCE: The Qin dynasty (followed by the longer-lasting Han dynasty) establishes China's first united empire.

212 BCE: Roman citizenship granted to all free inhabitants of the empire.

100 BCE: Rome becomes the largest city in the world.

150–650 CE: Teotihuacan becomes the largest city in the pre-Columbian Americas, with a peak of 125,000 inhabitants.

300–1200 CE: Ghana empire established in modern-day Mauritania and Mali.

410: Visigoths sack Rome; the final fall of the western Roman empire follows in 476.

661–750: The Umayyad caliphate is the largest empire in terms of area up to this date, stretching from modern-day Georgia, Uzbekistan and Pakistan across the Arabian peninsula and North Africa and up into Spain and Portugal.

1055: Seljuk Turks capture Baghdad.

***c.* 1200:** Incas settle in Andean valley in Peru.

1200–1400: Mississippian Culture in North America reaches its peak, with extensive areas along the Mississippi River under cultivation, and towns of up to 20,000 people.

1206: Foundation of sultanate of Delhi.

1211: Mongols begin conquests of northern China and across Eurasia.

1368: Ming dynasty established in China.

1393: Timur (aka Tamerlane) sacks Baghdad.

1405: Beginning of Zheng He's voyages in Indian Ocean.

1438: Beginning of period of Inca conquests.

EARLY TRADE ROUTES

Humans were trading long before they became settled farmers, exchanging such items as tools and decorative objects. Trading allows those with a surplus of certain goods to obtain other goods they desire, and it also promotes the transfer of cultures and ideas.

Once farming began, producers and consumers would sell and buy surplus agricultural produce at local markets. But long-distance trade is more complex: it calls for middlemen – merchants – prepared to accept risk and delay in anticipation of a profit further down the line.

One of the earliest known long-distance trades was in obsidian. This hard black volcanic glass was one of the most valued materials before the development of metallurgy, because it can be worked to make very sharp blades. As early as 14,000 BCE, obsidian was being traded into the Levant and northern Mesopotamia from sources in Anatolia (Asiatic Turkey).

Several millennia later, from around 3000 BCE, the ancient Egyptians were importing ivory from southern Nubia, copper and turquoise from Sinai, and amber from the Baltic. By the 1st century CE trade networks linked China, Japan, South-East Asia, the Indian subcontinent, Central Asia, Arabia, East Africa and the Roman empire. By the 2nd century CE, the population of Rome itself may have exceeded a million. The supply of goods to support the city was a spectacular economic, governmental and logistical achievement, with grain imported in bulk from Sicily, Tunisia and Egypt.

In the ancient world, and right up to the early modern period, the great powerhouse of long-distance trade was Asia. Spices went by sea from South-East Asia, while the precious fragrances frankincense and myrrh followed the Incense Route by land and sea from southern Arabia to the Mediterranean and beyond. The merchants of southern Arabia also

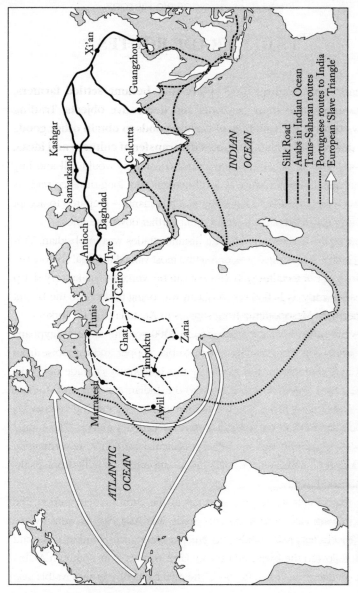

The Silk Road and other major historical trade routes

used the Incense Route to sell on gold, ivory, pearls, precious stones, spices and textiles that reached their ports from Africa, India and the Far East. Also prominent from the 8th century CE were the trans-Saharan caravan routes that carried gold and ivory from West Africa up to the Mediterranean. Various African kingdoms grew rich and powerful on this trade. In the Sahel belt of West Africa and along the coast of East Africa, the spread of Islam and the growth of trade were linked to the expansion of cities, such as Timbuktu on the River Niger, Kano in northern Nigeria, and Mogadishu and Mombasa on the Indian Ocean.

The most important long-distance trading route, from the 1st century BCE, was the Silk Road – actually a number of different overland routes from China through Central Asia and the Middle East to the Mediterranean Sea and beyond. Silk was one of China's chief exports, and in great demand by wealthy Roman women. Later, porcelain also became a key export. In the reverse direction the Chinese imported gold, silver, precious stones, ivory and other natural rarities.

At this stage China was technologically much more advanced than Europe. It was along the Silk Road that the Chinese inventions of paper, printing, gunpowder and the magnetic compass reached the West. Diseases also moved westward, including the plague that arrived in 542 CE in Constantinople, capital of the Eastern Roman (later Byzantine) empire. The closure of the western end of the Silk Road after the Ottoman Turks captured Constantinople in 1453 spurred countries such as Portugal on the Atlantic seaboard to establish new maritime trading routes to Asia round the southern tip of Africa.

'The Chinese porcelain . . . is exported to India and other countries, even reaching as far as our own lands in the West.'

Ibn Battuta (1304–68), Muslim traveller and observer

THE BIRTH OF CITIES

The development of trade and the shift to crop cultivation encouraged the growth of cities. Cities grew to facilitate the complex webs of exchange, while the production of regular food surpluses enabled some workers to specialize in other tasks. Urban development relied on agrarian systems that were able to support large populations. The richest agricultural areas often lay in fertile river valleys, such as the Euphrates and Tigris of Mesopotamia (modern-day Iraq), the Nile of Egypt, the Indus of modern-day Pakistan and the Yellow River of China. These were the sites of some of the earliest towns and cities.

Some early cities started as focuses of spiritual power, and only later became associated with learning, culture and law. For example, in the Mesopotamian cities of Nippur (first settled in around 5000 BCE) and Uruk (developed in about 3500 BCE), the most important feature was a sacred enclave consisting of a raised mud-brick ziggurat temple complex. The priests not only represented divine power, but also administered much of the city's land, recording and storing agricultural production. The ruling elites of these cities often also occupied buildings within the religious precincts. Later, many cities, including Jerusalem, were laid out around a central temple, as were the Mayan cities of Mesoamerica such as Chichen Itza, whose temple pyramid (El Castillo) still dominates the site.

In ancient China, a strong economy, based on the production of millet and rice combined with a sophisticated administrative system, ensured that the state could support a large urban population. Under the Shang dynasty (*c.* 1800–1027 BCE) there were a number of capital cities. The Zhou dynasty (1027–403 BCE) that succeeded the Shang again had a number of capitals, and it is this dynasty that offers the earliest documentary evidence we have of city planning. The principles of Zhou urban design, which continued to underpin Chinese grid layouts into the modern era, were based upon a holy square system derived from a mixture of cosmology, astrology, geomancy and numerology.

Abandoned cities

Cities did not only expand: sometimes they shrank or were even abandoned. Angkor, in present-day Cambodia, was one of the world's largest pre-industrial cities, its magnificent Angkor Wat temple one of thousands of religious buildings. It was mostly abandoned during the 15th century, for reasons unknown, though historians have offered explanations ranging from invasion to plague. Alternatively, crucial features of infrastructure, such as irrigation systems, may have broken down, leading to food shortages. Teotihuacan (in modern-day Mexico) was the largest city in the pre-Columbian Americas during the first half of the 1st millennium CE, but in 550 CE its main buildings and monuments were attacked and burned, possibly by invaders, but possibly during a violent internal uprising against the ruling class.

Under the Tang dynasty (618–970 CE), Chang'an (modern-day Xi'an) was the capital and had a population of about 2 million. The city's symmetrical layout was used to organize specialized and orderly functional neighbourhoods whose demarcation stemmed from deep-rooted Chinese ideas about the spiritual efficacy of spatial arrangements and alignments. Tang China sustained more than ten cities with populations of over 300,000. During the later Song dynasty (960–1279), the mercantile hub and metropolis of Hangzhou had 1 million residents at a time when London numbered only 15,000.

For the sake of security and control, early cities were often walled. The Greek historian Herodotus claimed that the walls of Babylon were 100 metres high and wide enough to allow two four-horse chariots

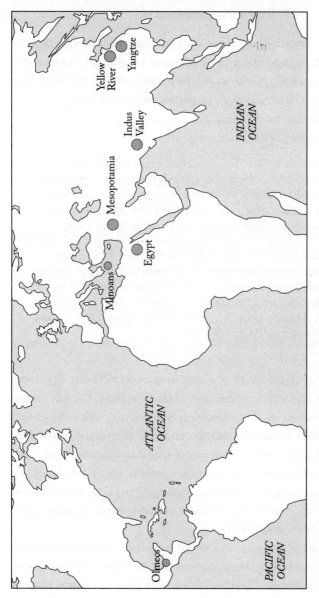

Early urban civilizations, from the 5th to the 2nd millennia BCE

to pass each other. Artisan quarters, zones where the exponents of a particular trade were gathered, were a common feature. For artisans such as blacksmiths, glass blowers or potters there were benefits in close proximity: they relied on the same raw materials and tools and offered clients a familiar location. Residential areas sometimes extended beyond the walls, while the walled area afforded asylum in times of strife.

TRANSPORT

Over the millennia, the growing importance of trade encouraged the development of new means of transport, which served first to connect, and then to integrate, distant regions.

A range of modes of transport served different environments. On land, the chief means were walking, riding and using draught animals to carry loads and to pull wagons. But water – whether rivers, lakes or sea – often offered faster routes than land (with its swamps, forests and broken ground), and often enabled greater quantities of goods to be transported. For land routes, key points were where rivers could be forded or bridged, or where passes pierced mountain ranges. Water transport called for navigable rivers and natural harbours, features that often dictated the location of human settlement. To this day many of the world's most important cities have grown on a navigable river or a natural harbour. Although transport technology improved over the centuries, its basis remained much the same – dependent on human or animal muscle power, wind, or the flow of water – until steam power transformed land and sea transport in the 19th century.

The driving of livestock for sale at often distant markets helped to establish overland routes. These long journeys, often of hundreds of kilometres, caused the animals to lose weight and thus value, so they would be fattened again by grazing near markets. All of this drove up the cost of meat, so that only the wealthy could usually afford it. The staging of lavish feasts in traditional stories from around the world highlights how rare and desirable meat was.

Before the steam engine, travel by water depended on climate and weather. Winter freezing, spring snow-melt floods and summer drought all caused problems, as did natural silting, which could make rivers too shallow to navigate. In many fast-flowing rivers, it was impossible to use horses on the banks to pull boats upstream against the current. As a result, for example on the River Rhône in France, boats were built upstream then, once they had travelled downstream with their loads, would be broken up for timber.

At sea, storms were always a danger, while contrary winds and currents hampered travel. The circular or counter-circular directions both of prevailing winds and of ocean currents affected the spread of seaborne migratory or trade routes. For example, it was difficult to sail the southern Pacific from east to west as the current pushed boats northwards. In the Atlantic, the routes established by European navigators by the 17th century exploited the prevailing winds and currents, heading southwestward from Europe towards the Americas, and returning in a north-easterly direction, several hundred miles north of the outward route. The hazards involved in navigating over long distances also affected travel on the oceans. While early mariners could navigate to some extent via Sun and stars, and by using the magnetic compass introduced to Europe from China in the later Middle Ages, it would be many more centuries before they could identify their longitudinal position.

On land, animals were a key alternative to human porterage (common, for example, in sub-Saharan Africa and in the Andes mountain chain of South America). A range of different mammals were enlisted as draught animals in different parts of the world – oxen, horses, asses, elephants, various members of the camel family (see p. 84). At first, draught animals would either just carry or drag their loads (often on some kind of primitive sledge). The wheel brought a major advance (see p. 86), but in most parts of the world the absence of good roads (roads were usually quagmires in winter, and grooved by ruts in summer), let alone an integrated road system, still restricted overland movement. In many places, the construction of canal networks – able to carry greater loads – preceded the establishment of well-maintained road systems.

FROM BARTER TO MONEY

The simplest way to trade goods is to barter. In barter, goods are exchanged and no money is involved. The goods may be raw materials such as grain, manufactured items such as pots, or services such as labour or story-telling.

But barter is a highly inflexible system, as it relies on each of two traders having exactly the goods that the other wants. There is little evidence that barter was ever used systematically in societies without money or credit. It seems more likely that once humans started to regularly trade goods they would soon have moved on to agreeing to keep a tally of goods exchanged so that the seller of a good could retain a 'credit' for later use.

Both money and credit arose for this reason. Credit, at its simplest, is a record of how much one party 'owes' to another. Money is anything that has a generally agreed value, and lets credits be transferred between parties. What matters is that sellers can go to market and ask for a certain sum for their goods, in the knowledge that they can use that sum to buy different goods. Money also makes it easier to store wealth, or to borrow it.

Where there was trust within a trading community, and no need to carry money over distance, a simple tally system or valueless tokens could be used. But where there was less trust, or trading over longer distances, other solutions were needed. One early record of credit was the tally stick, a notched stick broken in half, with the creditor and debtor taking one half each (since no two sticks break the same way, it was easy to check that there was no foul play). Some societies also chose to use objects with some intrinsic value or rarity for money, since items with intrinsic value could be used even if not accepted as money,

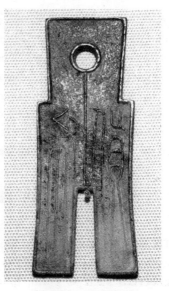

Chinese bronze hoe coin (early 1st century CE)

and rare objects were harder to fake. These might be tokens made from precious metals such as silver and gold, but there was a variety of earlier forms, including rare seashells (valued for their use in jewellery), useful tools, wheat and cattle.

Shell money has been used at various times on virtually every continent. Cowrie shells served as tokens of exchange around the Indian Ocean as early as 1200 BCE, and shell money was still legal tender in parts of West Africa through to the mid-19th century. In China, cowries were such important tokens of exchange that the classical Chinese character for 'money' or 'currency' derives from a pictograph of a cowrie shell. As time went on, people living some way from the coast could not obtain enough cowries for their trading needs, so began to make token versions out of available materials such as horn, bone, stone, clay, bronze, silver and gold.

Gold coins of Croesus, King of Lydia (seventh century BCE) and a silver Macedonian Tetradrachm depicting the head of Apollo (5th to 4th century BCE)

Around 1100 BCE the Chinese had adopted another system of tokens. These were miniature replicas of tools and weapons – previously valuable barter items – cast in bronze. But the sharp points of miniature hoes, spades, daggers and arrows made for awkward handling, and as time went by these items came to be represented by metal discs.

The first 'proper' coins, however, were minted on the other side of Asia, around 560 BCE in the kingdom of Lydia, in what is now Turkey. They were made from a mixture of gold and silver called electrum, and were stamped with the seal of the king, as a guarantee of value. As the metalworkers advanced in skill, they added more details, to demonstrate that each coin had the same metal content and the same weight. What counted about money, they realized, was that people must trust it.

Egyptian wheat money

The intrinsic value of a monetary token is rarely as high as its exchange value. Cowrie shells and gold and silver have a certain value, based on their decorative possibilities, but their use as currency enhances their value. Currencies that use everyday commodities are more rare. A notable exception was the wheat currency of the ancient Egyptians. For millennia they based a complex banking and financial system on wheat, which, as a staple food for the whole population, had an instant and significant intrinsic value. In many parts of the world, such a monetary system would not be possible, because of the unpredictability of harvests, but the annual flood and the dependable soil of the Nile valley meant that the Egyptians could rely on wheat as a stable – if bulky – currency.

PAPER MONEY

For two millennia after the Lydians first minted coins, these small discs made from precious or semiprecious metals were the commonest forms of money in many parts of the world. But for large transactions, coins are heavy and bulky, and paper banknotes, promising to pay the bearer on demand, in the end proved much more convenient.

The first such promissory notes were produced in China in the second century BCE and were made of leather. After the Chinese invented paper, they quickly realized it was the ideal material for banknotes, and paper notes began to circulate locally in the seventh century CE, and more generally in the tenth century.

In the later Middle Ages merchants in Italy and Flanders began using personal promissory notes, and these IOUs became payable in coin to anyone who had them in their possession. It was not until the 1660s that the first European banknotes were printed. Initially notes were printed by banks or other private institutions, although later this role became the sole prerogative of national governments.

The first European government issue of paper money actually took place in the North American colonies. Because coin coming from Europe by ship could take weeks if not months to arrive, colonial governments sometimes had to resort to issuing IOUs. The first instance was in French Canada in 1685, when the governor denominated and signed playing cards so they could be used as cash.

Through the 18th century paper money helped to expand trade, especially internationally, and banks and wealthy merchants began to buy and sell foreign currencies, so creating the first currency markets. If traders believed the government of a particular country was strong and stable, its currency was likely to increase in value against other currencies. International competition led countries to try to affect the value of a rival's currency, either by pushing up its value so its goods became too expensive to export, or by pushing it down and reducing the rival's ability to import goods – and to pay for war.

'All these pieces of paper are issued with as much solemnity and authority as if they were pure gold and silver . . . Kublai Khan makes them to pass current universally over all his kingdoms and provinces and territories.'

The Travels of Marco Polo (1298)

CREDIT, DEBT AND INVESTMENT

Even some of the earliest economies had systems of credit. And wherever there are systems of credit there are periodic debt crises. Many wars and revolutions have had their roots in problems of debt and debt remission.

Tally sticks and similar records of credit trace back at least as far as ancient Mesopotamia – as do offers of interest-bearing loans. Wherever such loans were available, there was always the chance that borrowers could fall into arrears and have their possessions seized, or, worse, have members of their family taken into slavery as 'debt peons'. To solve the social problems this caused, and to retain their subjects' loyalty, Sumerian and Babylonian kings announced periodic 'jubilees' in which consumer debts were annulled. Later Greek and Roman leaders responded to threats of debtor revolt by issuing payments to citizens, for instance through negative tax rates.

Knowledge of the problems debt could cause was also reflected in moral pronouncements, whether religiously sanctioned or not. In *The Republic* the Greek philosopher Plato put the rhetorical question: 'Is not justice just a matter of paying one's debts?' – to which the answer was that justice is far more complex than that. Usury (the issue of interest-bearing loans) was banned by the medieval Christian Church. At the same time, Islam permitted only profit-sharing loans (where the lender could claim only a share of profits from the debtors' investment, not interest) and banned debt slavery. But the banking systems that developed in Europe (starting in Italy in the later Middle Ages) and that underwrote a great expansion of commercial activity were based on lending money for profit.

It can furthermore be argued that without the relaxation of laws on usury in England in 1545, the Industrial Revolution that got under way two centuries later (see p. 164) would simply not have happened: the lure of a return on capital encourages people to risk their money by investing it. And so the prospect of income through lending would become a key feature of the capitalist system as it has since developed. The foundation of the Bank of Amsterdam in 1609 (precursor of all modern central banks) and of the Bank of England in 1694 gave the Netherlands and Britain more sophisticated and stable credit systems than existed anywhere else, and both countries saw significant growth in their economies.

'Annual income twenty pounds, annual expenditure nineteen pounds nineteen and six, result happiness. Annual income twenty pounds, annual expenditure twenty pounds nought and six, result misery.'

Mr Micawber, a character in Charles Dickens's novel *David Copperfield* (1850). Like Dickens's own father and thousands of others, Micawber was put in a debtors' prison for failing to repay his creditors

The increasing use of paper money encouraged further growth in debt issuance. It made it easier to borrow, but also harder to value a borrower's assets and thus to gauge how much that person should be able to borrow at an acceptable level of risk. This situation was impaired by the unpredictability of the economy. Factors like storms and disease could wreck production and trade. Nor could such misfortunes be adequately offset by insurance, which was another feature of the economic system that gradually grew over time.

The combination of paper money and debt issuance means that liquidity – money available to lend – cannot always be relied on. Changes in credit availability, sometimes in the shape of cycles of credit expansion and contraction, create patterns of economic booms and busts and lead to the differing problems posed by high rates of inflation and deflation. For example, in France, speculation in share values and a massive spurt in the amount of paper money led to a financial collapse in 1720 in which banknotes were no longer valid: the system of paper currency had collapsed.

The underlying causes of economic recessions and depressions in the modern world are the same as those that affected the early economies of Mesopotamia, and it may well be that issues of debt and debt forgiveness at both local and international levels will continue to weigh on the political and economic choices of the future.

WRITING

There can be little doubt that writing has been the most important tool in humankind's intellectual development. Before the development of writing, the accumulated knowledge and experience of an individual or a community could only be passed on orally, inhibiting its volume and variety. Knowledge could be lost with the death of an individual, or distorted by the flaws of human memory.

Once there was writing, knowledge could be recorded and stored over time. Once books and libraries existed, people no longer had to rely on memory, and could potentially access the accumulated wisdom of the ages.

Attempts to leave permanent records can be traced back at least some 20,000 years, during the last ice age in Europe, when Palaeolithic hunters carved regular groups of incisions in bones and antlers that may have served a calendrical function, perhaps recording the migratory movements of prey animals such as reindeer.

But true writing has to be far more flexible. The written symbols have to convey the actual words and sounds of a spoken language, not just broad ideas. A single writing system may be used to represent a number of different languages. For example, the Roman alphabet, which is about 2,500 years old, is used to write a number of European languages, from Romanian to Norwegian.

Different writing systems arose independently in places as diverse as the Near East, Mesoamerica, the Indus valley and China. The earliest was that developed around 3100 BCE in Mesopotamia. This was cuneiform, a word meaning 'wedge-shaped', referring to the marks the scribes incised with a stylus in their clay tablets.

Different writing systems

There are three main types of writing. In logographic scripts, such as Chinese, each symbol stands for a whole word. In syllabic systems, such as ancient Babylonian cuneiform and Japanese, each symbol represents a single syllable. In alphabetic systems, such as Greek, Hebrew, Arabic and our own Roman system, each symbol generally represents a single sound.

Writing first arose in early urban societies, which were more stratified than in pre-urban times. The ruling elite needed it as a means to keep control over masses of commodities and numerous subjects. The monumental inscription of a ruler's name – as found for example on stone slabs in Mesoamerica and on ancient Chinese oracle bones – helped to reinforce his or her unique and powerful status.

In due course most (but not all) literate societies began to use writing for a wider variety of purposes: business contracts, letters, religious rituals and laws, both religious and non-religious. Written literature arrived later: for example the Mesopotamian *Epic of Gilgamesh*, originally transmitted orally, was not written down until the 7th century BCE.

While we talk about the earliest 'literate societies', there were in fact very few people who could read and write. Those who could were usually trained scribes; even the rulers, the chief beneficiaries of writing, may often have been illiterate. Indeed, all through the ancient world and through the Middle Ages, literacy was restricted to only very small minorities. It was not until the coming of printing that the benefits of writing began to spread more widely (see p. 154).

LAW

Law, originally imposed by family and community leaders in the form of taboos and obligations, became more complex as societies developed. As a formal system, law was linked to the spread of both government and writing.

There is no way to establish to what degree concepts of right and wrong are innate, evolutionary, or developed within society. People argue either that they developed as part of religious practice or that they pre-date religion. But it is clear that by the time that the first laws were written, they served in part to record what a society did and did not judge permissible, and that this varied with the needs and values of the society. Laws were also a codification of those values that leaders or governments wanted to impose upon their subjects, but over time they would also come to be a restraint on the absolute power of the government. In ancient Athens, for example, the Draconian constitution, or Draco's code, was created in response to the unfair and arbitrary modification of oral laws by the aristocracy in the 7th century BCE. The Magna Carta (1215) marked a parallel moment in English history because it set limits on royal power.

Legal codes have varied as to whether they reckoned a criminal act to be an offence against the individual or community, or else against the state. This connects to the distinction between retributive justice (where the law defines how a lawbreaker should be punished) and restorative justice (where lawbreakers must make good the harm they have done). Many early legal codes incorporated a strong retributive streak. For instance the code of the Babylonian king Hammurabi (dating from 1754 BCE, one of the earliest legal codes that can still be studied in detail) ordains 'an eye for an eye, a tooth for a tooth', in wording very similar to the law of Moses in the Old Testament. However it also required compensation for crimes of property, as did both the earlier Code of Ur-Nammu in Sumer, and the Jewish Pentateuch. In Rome the Law of the Twelve Tables (450 BCE) required a convicted thief to repay twice the value of the goods they had stolen.

Laws didn't deal exclusively with crime and punishment. They also provided peaceful means of settling disputes and (in those societies that recognized private ownership) as a means to record the ownership of property. They also included rules for how contracts and trades should be regulated.

'At his best, man is the noblest of all animals; separated from law and justice he is the worst.'

Aristotle, *Politics*, Book 1 (4th century BCE)

Legal codes have also varied in their balance between protecting the rights of citizens, and imposing duties. For instance, protecting free speech or religious practice has been an important part of the law in some societies, while in others the law has included religious requirements or banned certain kinds of public discourse, such as criticism of the government, or statements regarded as blasphemous. Some codes have specified duties such as fixed terms a citizen must serve in the army.

And many legal codes have included laws that restrict personal freedom on specifically moral or religious grounds. For instance homosexuality has been (and remains) illegal in many countries, no matter if practised between consenting adults. The tension between protecting rights and imposing responsibilities on citizens also relates to arguments about the legitimacy of government itself (see p. 168).

Such debates continue. Current legal systems include both retributive and restorative elements, and countries worldwide vary as to how they balance rights and responsibilities, and how authoritarian their approach to lawmaking is.

ANCIENT EMPIRES

When states expand and begin to rule over a number of peoples, they can be regarded as empires. Such empires first developed in the more settled zones of the world, largely because increased wealth provided the means for their growth.

Egypt, the Middle East, northern India and China were the sites of some of the earliest empires. They were headed by rulers who claimed authority as the representatives of gods. The first empire in Western Asia was founded in about 2300 BCE by Sargon, who united the city-states of Sumer (southern Iraq) and conquered neighbouring regions of Mesopotamia. An empire based on the city of Ur followed, and later the Babylonian empire of Hammurabi (1790–1750 BCE). In China both the Shang dynasty (1600 BCE) and its successor the Zhou dynasty (1100 BCE) ruled over larger areas than their contemporary Near Eastern empires. The Zhou broke up into many smaller states in 770 BCE, and this situation persisted until the Qin dynasty united China in 221 BCE.

'An empire founded by war has to maintain itself by war.'

Montesquieu, *Considerations on the Causes of the Grandeur and Decadence of the Romans* (1734)

Military strength was crucial to imperial expansion. For instance the Neo-Assyrian empire (911–612 BCE) was able to conquer much of Mesopotamia, Anatolia, the Levant and Egypt partly because its professionally trained soldiers wielded iron weapons. Military force was decisive in creating and expanding the Persian empire in the 5th century BCE, the Macedonian empire in the 4th century BCE, and most famously the empire of Rome, which by the end of the 1st century CE stretched from Britain to Egypt and Syria.

While military force was required to create an empire, keeping it going was a more complex matter, which relied on government, bureaucracy, economic measures and more. In particular, empires needed to keep the support of conquered areas. (This was easier than in the modern world, as ideas of national self-determination, let alone democracy, were absent.) One means of doing so was to incorporate the conquered peoples. This might be done by spreading the religion of the empire, or by tolerating and co-opting local religions. In some cases, such as Rome, the empire might offer the right of citizenship to some of the local population to win their loyalty.

Empires also relied on the transport of resources, both by protecting trade and by imposing taxes. Strong empires provided security and stability in a world that had little of either, and in some cases this would win them short-term popular support. While the Romans levied tax, seized slaves and imported goods from the peripheries of their empire, they also believed they had a 'civilizing mission' – a concept first promoted by writers such as Cicero. Across the empire local populations started to build towns and roads in the Roman style, and adopted Roman customs in food, clothing and horticulture. In turn the empire recruited local administrators, and over time the Romans also intermarried with the populace. (Alexander the Great had his own ingenious short cut to stability: he ordered his generals to marry into the local ruling class. At Susa in 324 BCE he had eighty of his generals married to Persian princesses from the defeated Achaemenid empire.)

In India, the Mauryan empire (321–185 BCE), the first to conquer the entire Indian peninsula, illustrates the challenges of imperial rule. In the reign of Asoka the Great (304–232 BCE) it has been claimed that the empire contained the world's largest city of the time (Pataliputra) and that the empire had an army of 600,000 infantry and 30,000 cavalry, as well as 9,000 war elephants. But it also relied on a complex administrative structure. Each province was ruled by a member of the royal family. Local rulers lasted so long as they raised and paid their taxes, but found their loyalty constantly under scrutiny by the royal representatives and by a network of spies. In return the empire built public facilities such as irrigation systems and roads. It also maintained a justice system and paid to clear and reclaim forest land for farming (a crucial task when economic prosperity relied on agriculture). Trade routes were fostered and good relations with trading partners maintained. The result was an extended period of peace and prosperity, which made the task of maintaining loyalty to the empire an easier one.

In China, the level of organization rose to still higher levels. Under the Qin dynasty the administration of the empire was carried out by a large civil service whose members were chosen at first from recommendations by local officials. The subsequent Han dynasty (206 BCE–220 CE) refined the system and established a university that taught civil servants Confucian principles of government (see p. 132). They were tested rigorously before joining the bureaucracy, which would be the bedrock on which many subsequent Chinese empires were based.

WHY EMPIRES FALL

For over half a millennium Rome spread its civilization by force of arms across much of Europe, North Africa and the Near East. Why then, in the 5th century CE, did Roman power collapse?

Rome had been under pressure for some time. Germanic tribes had been pushing at the frontiers of the empire for centuries. They in turn were coming under pressure from warlike peoples spreading out of Asia, such as the Huns. The Roman empire relied more and more on the army, and this encouraged a succession of generals to declare themselves emperor. The result was political instability and civil war. The cost of defending the empire led to crippling taxation and inflation. Trade and agriculture suffered, and famine and epidemics further damaged the fabric of society. In the 4th century the seat of Roman power moved to a new capital in the east, Constantinople (now Istanbul). Rome itself was left to its fate. In 410 CE it was sacked by the Visigoths, and the last Roman emperor in the west fell in 476.

Some empires have ended with catastrophic speed. Alexander the Great of Macedon destroyed the Persian empire in the 330s BCE. The Aztecs in Mexico and the Incas in South America were extinguished within two or three years of the arrival of the Spanish in the early 16th century (see p. 122).

Over some 7,500 years, successive ruling dynasties in China broke down for various reasons – internal rivalry, peasant revolts, foreign invasion – but each new dynasty from the Qin onwards inherited both the imperial territory and the administrative power structure of its predecessor, and thus China remained as a united empire throughout most of its history. Contrast this with the end of the Mauryan empire in India, which failed to bequeath such robust imperial institutions. This is partly because the

Chinese system became meritocratic, and capable individuals ran the empire even under weak or unpopular emperors, whereas, after the death of the popular and charismatic Asoka the Great in 232 BCE, the Mauryan empire went into decline and had fallen within fifty years.

The reasons why empires fall tend to be related to the challenges all empires face. To keep local populations loyal, maintain military strength, encourage economic prosperity and build an administrative structure strong and flexible enough to rule distant territories are all difficult challenges, and failure to achieve any one of these can lead to disintegration or collapse.

POLYTHEISM AND MONOTHEISM

Humans may have practised various forms of religion for tens of thousands of years (see p. 71), but we know little about what they believed until the advent of literate societies, in the 3rd and 2nd millennia BCE.

Different states had different mythologies and cosmologies, often serving to explain and justify the power of the ruler. Most early religions were polytheistic (had several gods), and, like the state and the family, the gods and goddesses were arranged into hierarchies. As well as mirroring cultural identity and the structures of power, these pantheons of gods and goddesses often also represented the elemental forces of nature and the cycles of life and death.

In ancient Egypt, for example, the Sun god and supreme judge was Ra, and the pharaoh, as 'Son of Ra', was accorded divine status. The pharaoh was also known as Horus, the falcon god, son of the goddess Isis and her husband Osiris, god of the dead, of vegetation and of the annual Nile inundation that watered the crops. Osiris was slain by his brother Seth, god of disorder, violence and storms, but Horus in turn vanquished Seth. All this served to underpin the rule of the pharaoh as the bringer of peace and prosperity to his land.

The twelve Olympian gods of the ancient Greeks likewise reflected power structures, human qualities and the forces of nature. For example, Zeus, the ruler of the gods, was also the god of the sky, while his wife Hera, to whom he was frequently unfaithful, was guardian of marriage and childbirth. The Greeks told many stories about their gods, who had more than their fair share of human weaknesses and follies – lust, pride, anger, jealousy. The Romans had their own counterparts to the Greek gods, and latterly the Roman emperors often assumed semi-divine status.

Hinduism emerged in India from around 1500 BCE. It too has a pantheon of gods, some of whom are married to each other, while some have a number of different 'avatars' – manifestations of themselves. Three of the major gods represent the cycle of life and death: Brahma is the creator, Vishnu the preserver, and Shiva the destroyer. Zoroastrianism, which emerged in Iran possibly around 1200 BCE, depicted life as a struggle between two gods, a 'wise lord' and a 'hostile spirit'. Assigning certain qualities, properties and powers to different gods and goddesses was one way that humans attempted to make sense of their own condition.

'I the Lord thy God am a jealous God . . .'

Exodus 20:5

When monotheism (belief in one god) first arose is not clear. Judaism, Christianity and Islam, all of which stress the battle between good and evil, trace their monotheism to the covenant between God and Abraham, supposedly around 2000 BCE. However, this story was not written down until some 1,500 years later, and it may have been only then that Yahweh became the sole god of the Jews, and later on of Christians and Muslims.

EPICS

Many early civilizations produced large-scale works of literature, often in verse, that embodied their own view of themselves. These epic poems usually focus on the adventures of one or more warrior-heroes, and are often intermingled with the myths and religions of the peoples who created them. For example, the Hindu epic of the *Ramayana*, composed in the 1st millennium BCE, tells of the lives, loves and battles of Rama, an avatar of the great god Vishnu.

The earliest epics were at first transmitted orally. The *Iliad*, attributed to Homer and first written down in the 8th century BCE, belongs to a long tradition of Greek oral poetry about a war between Greeks and Trojans that may have taken place several hundred years earlier. Similarly, the ancient Mesopotamian epic of *Gilgamesh*, written down in the 7th century BCE, recorded older tales that had been passed down, including the story of a great flood that also appears in the Jewish Bible and other myths and legends of the Near East. The earliest manuscript of the Anglo-Saxon epic *Beowulf* dates from the late 10th century CE, but the poem itself was probably composed some 300 years earlier, and tells a story mixing fiction and some historical fact, set in 5th-century Scandinavia and northern Germany, from where the ancestors of the Anglo-Saxons had migrated to England.

Epics are often about the origins of the people who created them, and their heroes and villains embody qualities regarded as desirable. The *Iliad* was the product of a warlike society and its hero Achilles has been described as little more than a highly effective killing machine. The eponymous hero of *Beowulf*, in contrast, although also a warrior and technically a pagan in the period in which the poem is set, is a much more nuanced character, and is given a range of Christian virtues

The Hindu god Rama, hero of the ancient Indian epic the Ramayana, *stalks the demon Marica, who has assumed the form of a golden deer*

to suit the values of the poem's audience.

Although early epics began in oral tradition and extended their cultural significance step by step, some later epics were created by their authors as self-consciously literary celebrations of a nation or culture. The outstanding early example is the *Aeneid*, by the Roman poet Virgil, effectively the official poet of the first Roman emperor, Augustus. Virgil builds on the stories of the *Iliad*, taking the Trojan prince Aeneas as his hero. Aeneas escapes from Troy as the Greeks burn it to the ground, and after many journeys and battles becomes the ancestor of the Romans. By tracing an ancestry back to an age when gods and heroes walked the Earth, Virgil endows both Rome and Augustus with legitimacy and dignity.

'Arms and the man I sing.'

Virgil, *Aeneid*, opening line (19 BCE)

WRITING HISTORY

What we now think of as history – the scholarly study of the past – emerged only slowly from the myths and legends people long told each other about their ancestors and origins. The first writings about the past often amounted to little more than king lists, tracing the descent of a current ruler, often back to a god, and lending divine sanction to the status quo.

Early epics such as Homer's *Iliad* may have contained faint echoes of distant historical events (see p. 125), but their purpose was to enthral and move their listeners, not to explain events, or to offer a balanced view of the past. Such grand heroic poems reflected and celebrated the values of the societies they came from. Histories began as foundation myths, the myths of peoples, dynasties and religions. Religion played a key role. Even to this day, historians cannot help but reflect the values and priorities of their own societies to some degree or another, although there is usually an attempt to strive for balance and objectivity.

The first to attempt to do this was Herodotus, a Greek who lived *c.* 485–425 BCE. In writing his account of the Greek–Persian Wars fought in the early part of his own century, he travelled round Greece, seeking out those who had actually taken part in the conflict. He states clearly that he cannot always vouch for the truth of a 'fact' – he is only reporting what has been related to him. He defined history as what can be truly discovered, rather than just stories about the past. Herodotus organized his source material systematically and critically, then formed it into a coherent and even-handed narrative.

'I write in the hope of hereby preserving from decay the remembrance of what men have done.'

Herodotus (*c.* 485–*c.* 425 BCE), *The Histories*

Herodotus' younger contemporary Thucydides (*c.* 460–*c.* 400 BCE) wrote about the Peloponnesian War between Athens and Sparta with similar even-handedness and with a stress on evidence, but many later classical historians quite plainly had agendas. The Roman historian Livy was a propagandist for the virtues of the (partly imagined) 'golden age' of the Roman republic just as the empire was superseding it. A century later his compatriot Tacitus emphasized the failings and vices of the emperors who had usurped the republic.

These authors embodied the concept of history as a source of examples to guide action in the here and now, a view that was explicit in the work of Plutarch (*c.* 46–*c.* 120 CE), a Greek biographer who took up Roman citizenship. In his *Parallel Lives*, he compared a series of famous Greeks and Romans, pointing out the ways in which they shared common moral failings and virtues. The character of leaders, Plutarch believed, was what shaped the destiny of the mass of humanity.

This 'great man' version of history, common for many centuries, not only in Europe but also in China and the Muslim world, has fuelled debate among historians. For instance, opposing Thomas Carlyle (whose own historic approach focused on 'heroes' of history) in the 19th century, Herbert Spencer argued that great leaders were merely the products of the societies they came from. Those who agree believe that history should focus on the broader causes of historical events, and attempt to detect deep underlying political, economic, social and cultural patterns and trends.

THE NATURE OF REALITY

Around two and half millennia ago, something remarkable happened in the history of thought. In various parts of the world, a number of individuals started to ask a fundamental question: what is real?

For hundreds of thousands of years, humans had been too busy surviving to bother with such questions. For them it was clear what was 'real': food, shelter, reproduction, the basic necessities of life. Religion (see pp. 71 and 123) had been around for a long time, but by and large the world of gods and spirits was inseparable from people's everyday lives. The two realms intersected and interacted.

In India from around 800 BCE there emerged the Hindu concept of *samsara*, or reincarnation – the idea that all creatures follow a cycle of birth, death and rebirth. Relief from the endless wheel of suffering in the material world was possible only for those individuals who had freed themselves from desire for earthly things, enabling them to achieve union with the divine. *Samsara* contrasts with the views of the monotheistic religions, in which each individual is born only once and dies only once, death being followed by a perpetual afterlife.

Samsara became key to the teachings of the Indian prince Gautama Buddha, who developed his ideas around 500 BCE. As in Hinduism, enlightenment is possible only if individuals recognize the illusory nature of the material world and free themselves from desire, whether for goods or persons or pleasure.

That the material world is illusory was also insisted upon by the Greek philosopher Plato (*c.* 427–347 BCE). He held that ultimate reality consists of 'forms' or 'ideas', of which individual material manifestations are poor copies. Natural things and virtues have perfect forms. In a famous parable, Plato describes some prisoners who have been chained in a cave, and can see only the wall in front of them. On this wall they can see the shadows of various objects cast by a fire, and take these shadows to be real. We are like these prisoners, Plato says, taking shadows for reality.

Later philosophers developed Plato's ideas, holding that reality does not exist independently of the mind. This position is known as idealism. Other philosophers – both in ancient Greece and later – have taken a very different view: materialism. Materialists hold that matter is the only reality, and that mind, ideas, emotions, and so on, arise out of the workings of matter. An early example was the atomic theory proposed by the Greek philosopher Democritus (see p. 134).

A third strand in Western philosophy is that of the dualists, who hold that mind and matter are both real, but are different in kind. A very different kind of dualism appears in Daoism, which emerged in China in the 4th century BCE. Daoism holds that the ultimate reality is the Dao ('the way'). The Dao underlies all events, and is also the flow of those events. It combines opposites: stillness and motion, good and evil, light and dark. The Dao has no being, is unfathomable and indescribable, and yet everything depends on it.

Change and motion

A question that exercised some ancient Greek philosophers was whether change and motion are real or illusory. Some held that reality is single and unchanging, while others insisted that reality is never fixed, but always in a state of flux.

WHAT IS THE GOOD LIFE?

Around the same time that various philosophers and religious thinkers started to question the nature of reality, a number of them also started to ask how people should lead their lives. It was one thing to tell people to obey the laws or face the consequences, quite another to ask what constituted the best way to live.

Some religions, such as Judaism and to a certain extent its offshoots, Christianity and Islam, take a legalistic and moralistic approach to the question. The holy books lay down the laws; infringe them and you will receive punishment in this world, and if not in this one, in the next. The core text here is the Ten Commandments, which, according to the Book of Exodus in the Old Testament of the Bible, God handed down to Moses. The good life according to these doctrines largely consists of not committing sin. Devotion to God also requires following certain rituals.

Early Hinduism took a similar approach: perform the religious rituals correctly, and after death you will dwell with the gods. With the concept of *samsara*, however (see p. 130), individuals could not just rely on ritual, but were urged to rid themselves of earthly desires, so as to break free from the eternal cycle of birth, death and rebirth. The Buddha built on this belief, teaching that attachment and desire only bring suffering. If you lead your life with the aim of gaining money, for example, or the goods that money can buy, you will never be content, any more than if your constant aim is love or pleasure.

Buddha urged his disciples to follow the 'middle way' between material self-indulgence and extreme asceticism. This 'golden mean' is a common thread through much thinking about what constitutes the good life. The Chinese sage Confucius (or Kongfuzi, 551–479 BCE) advised such a balance. He taught that to achieve happiness people must follow the 'will of Heaven' by cultivating five qualities: loyalty to the state, family love, due respect, kindness to strangers, and reciprocity between friends. Confucianism, with its emphasis on the stability of the family and state, provided the ideological bedrock for imperial China for millennia.

'Do not do to others what you do not want them to do to you.'

Confucius. This is an early formulation of what became known as 'the Golden Rule' of conduct, also enunciated by Jesus in Luke 6:31.

The golden mean urges moderation in all things. Among the ancient Greek philosophers, Socrates (as reported by Plato) applied it to education, advising a balance between gymnastics and music, as the former on its own he maintained breeds hardness, and the latter softness. Plato's pupil Aristotle defined virtue as using reason to identify the mean 'between two vices, one of excess and one of deficiency'. Aristotle gave as an example the virtue of courage, which lies between foolhardiness (an excess of courage) and cowardice (a deficiency of courage).

Plato wrote that the moral goal is 'becoming like God'. Aristotle believed that happiness was the greatest good, and emerged from leading a good life. Other Greek philosophers held that happiness is the greatest good and thus the main goal of life. Some associated happiness with pleasure, but insisted that reason and self-control were essential in achieving the greatest pleasure. Others defined pleasure as the absence of pain, either mental or physical, and suggested that this could best be achieved by serene detachment from desire.

For many of these thinkers, the good life in the personal sphere was echoed in the social and political spheres, where moderation of passion and the balancing of conflicting interests were essential to achieve harmony.

THE BEGINNINGS OF SCIENCE

Humans have long sought explanations of the physical world. At first this was a central role of myth and religion. Nearly all cultures have origin myths, telling how the world and the first people came to be. Other myths account for all sorts of natural phenomena.

The systematic observations made by many cultures of astronomical events such as equinoxes, solstices and lunar cycles had a more practical purpose: they related to the changing seasons, vital to both hunters and farmers. A striking example is the great Neolithic ceremonial mound at Newgrange in Ireland, built around 3200 BCE, whose inner chamber is only lit by the Sun on the winter solstice. To build such a structure, which could precisely mark the passing of another year, took accurate measurement and careful observation – two of the hallmarks of what we now call science.

At around the same time, the Sumerians developed a basic lunar calendar, while by 2500 BCE the Egyptians were using a solar calendar – essential for anticipating the annual flood of the Nile. The Sumerians introduced a sexagesimal number system (using base 60, probably because 60 has 12 factors; our system uses base 10, which has only four factors, 1, 2, 5 and 10). The Babylonians followed them, and instituted the hour as a unit of time, calibrated into 60 minutes, which meant that it was evenly divisible into spans of 60, 30, 20, 15, 12, 10, 6, 5, 4, 3, 2 or 1 minute. The Babylonians were great astronomers, and by 1500 BCE were using mathematics to plot the positions of stars and planets, and to predict eclipses. The ancient Egyptians took mathematics further, and evolved some abstract geometrical principles out of the practice of land surveying.

But in the ancient world it was above all the Greeks who laid the foundations of the discipline we now call science. Their philosophers broke ground in exploring not only the nature of reality and what the good life consists of (see p. 131), but also the facts of the physical world. Rejecting mythological explanations, they sought to find a single underlying principle.

For Pythagoras and his followers in the 6th century BCE this principle was number. They established that the Earth is spherical, and recognized that harmonies in music are based on numerical ratios. In the following century Democritus theorized that everything consists of tiny indivisible particles he called atoms, while Empedocles proposed that matter is made up of four elements: earth, water, air and fire. The concept of the four elements was taken up in the 4th century BCE by Aristotle. He set out to observe and catalogue a vast range of natural phenomena, biological and non-biological, and from these findings to derive more general truths – the basis of the scientific method.

Around 300 BCE Euclid laid out the principles of geometry, and Archimedes later pioneered both mechanics and hydrostatics. In the 2nd century BCE Aristarchus of Samos showed that the Earth rotates on its own axis and orbits the Sun, while Eratosthenes (*c.* 276–194 BCE), the Greek astronomer who became the chief librarian at Alexandria in Egypt, worked out the Earth's circumference with remarkable accuracy. Also in Alexandria, by then part of the Roman empire, and a major centre of intellectual life, Ptolemy (Claudius Ptolemaeus, *c.* 90–*c.* 168 CE) drew up a world gazetteer that included an estimate of geographical coordinates.

'By convention there is colour, by convention sweetness, by convention bitterness, but in reality there are atoms and space.'

Democritus (*c.* 460–*c.* 370 BCE), fragment 125

The Romans added relatively little to the Greek scientific heritage, and, after Rome's fall, Greek science was largely lost in Europe. But Muslim scholars kept it alive in intellectual centres from Córdoba in the west to Delhi in the east. They also made numerous innovations. Muslim scholars adopted the Indian idea of a place value system in mathematics, including the concept of zero. This is the origin of the 'Arabic' numeral system we use today – much more useful for difficult calculations than the systems used by either Greeks or Romans. However, it was not until the 17th century that modern science really began, with the so-called Scientific Revolution (see p. 156).

DISEASE PANDEMICS

One of the consequences of humans living closely together in cities was the way infections found it easier to take a hold on a population, passing rapidly from person to person. And the expansion of long-distance trade meant that disease could travel from one corner of the world to the other, a trend that has speeded up over time, especially since the spread of passenger aviation.

In some populations, certain diseases are endemic. Societies tend to adapt to them – they become part of everyday life and death. But sometimes a given disease will explode into an epidemic, and if it then spreads more widely, across frontiers and continents, it may turn into a pandemic. Such events can have dramatic impacts on the course of human history.

Medical historians do not like to make broad retrospective diagnoses of diseases described in ancient, medieval or early modern sources. The word 'plague' has been a blanket term for various different lethal epidemics, such as the plagues mentioned in the Old Testament, and the Plague of Athens of 430–427 BCE. The first case that *may* have been bubonic plague (caused by the bacterium *Yersinia pestis*, spread by rats and fleas, and characterized by buboes, black swellings in armpit and groin) was the Plague of Justinian in 541–4 CE. This epidemic spread all around the Mediterranean, killing perhaps one-quarter of the region's population. It hit at a time when it looked possible that the strength of the old Roman empire might be restored. Although the western Roman empire had been overrun by Germanic tribes in the previous century, the eastern emperor, Justinian, had embarked on an ambitious campaign of reconquest. But the chaos and devastation inflicted by the plague shattered all dreams of reunification.

The Black Death that spread from Asia in the 14th century, killing perhaps one-third of the population of Europe, was probably a mix of bubonic, pneumonic and septicaemic plague. The Black Death was a turning point in European social, economic and intellectual history. With so much of the agricultural workforce dead, those who survived could demand higher wages. The landowners put up resistance, and this provoked peasant revolts. Many took the Black Death as a sign of God's displeasure, both with his people and his Church. Various groups consequently questioned papal authority, anticipating the Protestant Reformation of the 16th century.

'Where are our dear friends now? Where are the beloved faces? . . . What tempest drowned them? What abyss swallowed them? There was a crowd of us, now we are almost alone.'

Petrarch, Italian poet of the 14th century, expresses the loneliness of the survivors of the Black Death

The Black Death spread to Europe from the Black Sea via ships of Genoese traders. In the 16th century the European 'voyages of discovery' to the New World caused similar devastation (see p. 161).

By the early 20th century there had been a revolution in our understanding of the causes of disease, but this could not prevent the emergence of new pandemics. Although outbreaks of influenza had occurred for centuries, the strain that swept the world in 1918–19, known as 'Spanish flu', caused unprecedented devastation. Estimates of mortality range from 50 million to 100 million, many more than the death toll of the entire First World War. Unlike other flu outbreaks, most of the victims were aged between twenty and forty, which maximized the demographic impact. Some scientists believe it is only a matter of time before humanity experiences another such pandemic.

EUROPE IN TRANSITION

After the fall of the last western Roman emperor in 476 CE, Roman imperial power persisted in Greece, the Balkans and Anatolia (Asiatic Turkey) in the form of the Byzantine empire. It steadily lost territory to a variety of invaders until the Ottoman Turks extinguished it in 1453.

But what about the power vacuum left in western Europe? At the start of the 5th century CE, Germanic tribes had overrun much of the area, and in the wake of the fall of Rome they had established a patchwork of kingdoms: Visigoths in Spain, Vandals in North Africa, Ostrogoths in Italy, Franks in Gaul (France) and western Germany, and Angles, Saxons and Jutes in England.

Although the Romans had regarded these peoples as 'barbarians', they soon became Christianized, as the Romans had before them. Although there was no central secular power, the western Church was unified under the Pope in Rome. By a process of conquest, towards the end of the 8th century the Frankish king Charlemagne succeeded in uniting France, Italy and much of Germany, and on Christmas Day 800 the Pope crowned him 'emperor of the west'.

But Charlemagne's empire was short-lived, and within a few decades of his death it had fragmented, while power passed to a mosaic of regional aristocracies. Western Europe came under pressure from a range of new invaders. In the east, Magyars from the steppes advanced into central Europe, and were only stopped by the German ruler Otto the Great at the decisive battle of Lechfeld in 955. Thereafter, the Magyars established their own kingdom in Hungary.

The people from the north

From the 9th century, various seafaring peoples from Scandinavia, known and feared as Vikings or Norsemen, traded, raided and established kingdoms from Russia in the east to the British Isles in the west. They colonized Iceland and Greenland, and even reached North America. One group settled in Normandy and adopted French culture. The Normans (the name comes from 'Norsemen') went on to conquer not only England, but also southern Italy and Sicily.

But the biggest impact on southern Europe came from a different direction: Arabia. In the early 7th century the Arabian prophet Muhammad had founded a new religion, Islam, and by the time of his death in 632 he had unified all of Arabia. He urged his followers to spread Islam further afield, and over the decades that followed Arab armies seized Byzantine territories right across North Africa, Syria and Palestine. By the mid-9th century the Arab empire extended from the borders of India in the east to the Iberian peninsula (Spain and Portugal) in the west. The spread of Islam was not just down to the military prowess of its followers. Some of the peoples in the Byzantine and Persian empires were tired of the religious persecution they suffered at the hands of their rulers, whereas many Muslim leaders showed greater toleration. The Islamic world of the Middle Ages also produced a range of philosophers, physicians, mathematicians and scientists, who built on and advanced the achievements of the ancient Greeks.

LAND, LABOUR AND POWER

From the early days of settled agricultural societies, control over land was a crucial feature of economics, society and politics in many parts of the world. Its legal and governmental basis varied greatly, but it was force that ultimately ruled.

In the Middle Ages in western Europe, the system now known as feudalism ruled from about the 9th to the 15th century CE. The greater nobles were granted control over their lands in return for military service to the king. Those of lesser rank, such as knights, held theirs in return for military service to the greater nobles. The peasantry farmed, but did not own, their small patches of land in exchange for work performed on the land of their lord. This system – serfdom – was gradually to break down as the payment demanded changed from labour and military service to money, although it persisted in parts of Europe, such as Russia, well into the 19th century.

Other societies around the world had similar systems. Control of the land entailed the control of society and was central to the unequal distribution of wealth. The specific form this took depended on environmental factors and the availability of labour. There were strong contrasts between systems of hierarchy and control where labour was in fairly short supply, as in Africa and eastern Europe, and where it was more plentiful, as in China, India, Japan and western Europe. The former stressed control over labour, the latter control over land. Labour shortages could lead to tighter control, but they could also allow for renegotiations of labour relations that gave more power to the peasantry. In western Europe it was partly the shortage of labour after the Black Death in the 14th century (see p. 136) that enabled peasants to call for money payments and helped to erode the old feudal system.

The labour services owed by the poorest in society varied around the world, depending on the type of economy. What stand out are the major contrasts between pastoral (animal husbandry) and arable (cultivation) systems. Pastoral systems imposed a less rigid work regime, partly owing to greater flexibility in terms of how territory and land was exploited. Arable systems played key roles in east Asia. In North America, different tribes had differing ways of life. Those who were pure hunter-gatherers had no sense of owning particular pieces of land, whereas those settled in arable farming communities had more permanent connections to particular territories.

Until the system collapsed in the mid-18th century, in the remoter Highlands and islands of Scotland the clans were also tribal, and based notionally on family groups, though their origins lay in a relatively lawless period in which local warlords offered protection to families in return for loyalty. Tribal systems tended to rely on a sense of belonging and sustained loyalty among their members, whereas in feudal systems this kind of blood linkage was mostly absent.

This was also true of slave systems, where outsiders (often captured during wars, or traded in special markets) could be forced to work for a particular owner. Slaves had few or no rights, and no property. They were viewed as the property of their owner, as were their children. This was an extreme version of labour control, and laid few or no duties on the slave owners. Slavery was widespread in the ancient world (for example in Egypt, Greece, Rome and China), the Islamic world, and also among the pre-Columbian civilizations of Mexico and South America.

Serfdom aside, in Europe slavery had largely died out by the 2nd millennium CE. However, once Europeans began to settle in the Americas from the 16th century, they set up extensive plantations to grow such crops as sugar, tobacco and cotton. To work these plantations, they imported millions of slaves from Africa, shipped in appalling conditions across the Atlantic. Many Europeans grew rich both from the slave trade and from the slave plantations, and their profits contributed to the capital that kick-started the Industrial Revolution (see p. 164).

The end of slavery?

Not until the 18th century did some people in Europe and North America, partly inspired by their religious faith, begin to campaign against slavery. The institution was gradually abolished in the northern states of the USA, while in 1807 Great Britain outlawed the slave trade in its empire, and slavery itself in 1833. But in the USA as a whole, slavery was not abolished until 1865, at the end of the Civil War between the slave-owning South and the slave-free North. It lasted much longer in some other regions, and although criminalized in all countries today, slavery persists in the shadows: the trafficking of human beings across borders, to be used forcibly as sex workers or domestic servants or agricultural labourers, is a global criminal activity.

CLASHES OF CIVILIZATIONS

The emergence of two aspiring world religions – Christianity and Islam – within a few hundred years of each other led to centuries of conflict, which many of the protagonists viewed in terms of good versus evil, light versus darkness. But was it really a battle of ideologies, or more simply a matter of vying for power and material possessions, using religion as a moral justification?

The Byzantine empire was the first part of Christendom to feel the military impact of Islam. In the 630s and 640s, shortly after the death of Muhammad, the Byzantines lost Egypt, North Africa, Palestine and Syria to the Arabs. Then in the 11th century new pressure came from the Muslim Seljuk Turks to the east, who destroyed the Byzantine army at the

The Genghis Khan Equestrian Statue, Mongolia

battle of Manzikert in 1071 and conquered Anatolia.

After Manzikert the Byzantine emperor appealed for help from the Christians of western Europe. Pope Urban II saw an opportunity to assert the primacy of the Roman Church over Eastern Orthodoxy. He preached an influential sermon at Clermont in France in 1095 that cited massacres by Muslims of Christian pilgrims to the Holy Land and called on the nobles of western Europe to free the Holy Land of Muslim control. Most of those who joined the First Crusade were Norman and

French nobles keen to seize land, loot and a place in heaven. In 1099 the Crusaders captured Jerusalem, massacred many of the mostly Muslim and Jewish inhabitants, and established a Crusader kingdom. As various factions struggled to keep control of the Holy Land, further crusades brought further bloody outcomes.

During the 13th century the Muslims regained control over the Holy Land, but they also faced a major new threat, this time from the east. At the start of the century, on the other side of Asia, Genghis Khan had succeeded in uniting all the tribes of Mongolia. Under his leadership, these horse-riding warriors conquered northern China, and then looked westward. After Genghis's death his successors continued to expand, into Russia and the Near East. In 1258 they captured Baghdad and killed the caliph. But the Mongol threat to Muslim rule in the Near East was ended two years later when they were defeated by the Mamelukes of Egypt.

The Mongols had more enduring success in China, where Genghis's grandson, Kublai Khan, overthrew the rulers of the Song dynasty and made himself emperor. In this case, however, rather than destroying the fabric of Chinese culture and society, Kublai Khan adopted Chinese ways, and so helped to preserve Chinese civilization.

'It may well be that the world from now until its end will not experience the like of it again'

Ibn al-Athir, *The Perfect History* (early 13th century), describing the campaigns of Genghis Khan in the Middle East

From the 14th century it was Christian Europe's turn to face an existential threat, this time from the Ottoman Turks. By 1402 the Ottomans had occupied much of the Balkans, and in 1453 they captured Constantinople, the last remaining major stronghold of the now defunct Byzantine empire. In the following century they conquered not only the Arab lands

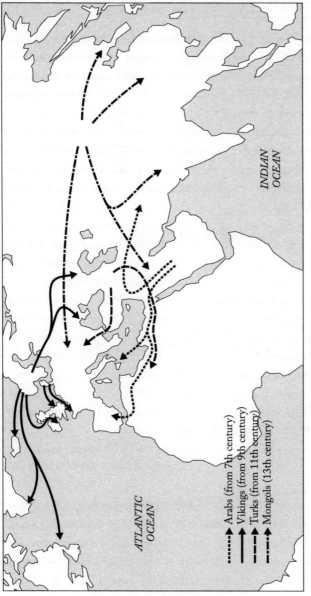

ATLANTIC
OCEAN

INDIAN
OCEAN

Arabs (from 7th century)
Vikings (from 9th century)
Turks (from 11th century)
Mongols (13th century)

Expansion and conquest

of North Africa, Arabia and the Near East, but also Hungary, and in 1529 stood at the gates of Vienna, then the capital of one of the great powers of Europe. Their failure to capture the city marked the end of Ottoman expansion into central Europe. In turn, the defeat of their fleet in 1571 at the Battle of Lepanto off Greece by the Holy League (an alliance of the southern European Catholic states) ended Turkish ambitions to extend their power into the western Mediterranean. After the Ottomans, in 1683, again failed to take Vienna, it was Christian Europe's turn to go once more on the offensive, and the power of the Muslim world began a slow process of decline.

PART FIVE

THE RISE OF THE WEST

Five hundred years ago Europe was something of a backwater. Since the fall of Rome a thousand years before, the continent had broken up into a jigsaw of small territories, often at war with each other. The real powerhouses – of intellectual inquiry, technology and trade – lay elsewhere, in China, India and the Muslim world. Across the Atlantic, in the Americas, there thrived civilizations undreamt of by Europeans. But from around 1450 the balance started to shift, as Europe began to assert itself on the world stage.

TIMELINE

1453: Ottoman Turks capture Constantinople.

1455: Gutenberg prints his first book.

1492: Columbus reaches West Indies. Last Muslim stronghold in Spain falls to Christians.

1498: Portuguese navigator Vasco da Gama reaches India via southern tip of Africa.

1517: Martin Luther starts Protestant Reformation.

1519-21: Spanish conquest of Aztec empire.

1519-22: First circumnavigation of the world, by Magellan and Del Cano.

1526: Beginning of Mughal conquest of India.

1532-5: Spanish conquest of Inca empire.

1543: Copernicus publishes his heliocentric theory.

1571: Holy League's victory at Lepanto ends Ottoman expansion in the Mediterranean.

1607: English begin permanent settlement of Virginia.

1644: Manchus (Qing) establish dynasty in China.

1648: End of Thirty Years' War fixes boundaries between Protestant and Catholic Europe.

1652: Dutch establish Cape Colony in southern Africa.

1683: Turks fail to capture Vienna.

1687: Newton publishes his law of gravity and three laws of motion.

1763: Seven Years' War ends with Britain dominant in India and North America.

1776: US Declaration of Independence from Britain. Publication of Adam Smith's *Wealth of Nations*.

1783: First manned balloon flight.

1785: Steam power first used in cotton mills.

1788: First British settlement in Australia.

1789: Start of French Revolution.

1792-1815: French Revolutionary and Napoleonic Wars.

1803: USA purchases a large swathe of North America from France.

1808-26: Spain loses most of its colonies in the Americas.

1825: First passenger steam railway, from Stockton to Darlington in England.

1830: Greece independent from Ottoman Turks.

1833: Abolition of slavery in British empire.

1844: First use of Morse's telegraph.

1848: Marx and Engels publish *Communist Manifesto*.

1848-9: Many failed revolutions in Europe.

1853: US fleet forces Japan to open up to Western trade.

1857: Indian revolt against British rule.

1859: Darwin publishes *Origin of Species*.

1861: Unification of Italy complete.

1861-5: American Civil War.

1868: Japan begins rapid modernization.

1869: Completion of first transcontinental railway line in USA. Opening of Suez Canal.

1871: Unification of Germany complete.

1876: Invention of telephone.

1884: At Berlin Conference, European powers carve up Africa.

1895: Marconi invents wireless telegraphy.

1903: First flight of heavier-than-air machine.

RENAISSANCE AND REFORMATION

Between the 14th and 16th centuries Europe underwent a cultural revolution. In philosophy and the arts, the Renaissance saw a reawakening of interest in the achievements of pre-Christian classical thinkers, artists and authors. In religion the Protestant Reformation questioned the authority of the Roman Catholic Church, and its claim to have sole care over people's souls.

For centuries learning in western Christendom had been confined to the monasteries, and was mostly concerned with theological issues. Most intellectual effort centred around matters of religious doctrine, while painting and architecture were also largely in the service of the Church, one of the wealthiest patrons.

The texts of the ancient Greek scientists and philosophers had been lost in the west until the 12th century, when Gerard of Cremona started translating Arabic versions of Greek texts into Latin. Subsquently writers such as Thomas Aquinas attempted to integrate Aristotelian philosophy into Christian theology – which was still regarded as the pinnacle of human intellectual achievement.

From the 14th century a group of Italian scholars, inspired by the poet Petrarch, proposed a new educational syllabus, based on classical literature, which they called *studia humanitatis*. This syllabus was to comprise five key subjects: rhetoric, poetry, grammar, history and moral philosophy. Theology played no part, although the humanists (as these scholars became known) did not go so far as to reject Christian doctrine. Instead they shifted the emphasis from debating how a person should serve God to asking how the virtuous man should act.

'After the darkness has been dispelled, our grandsons will be able to walk back into the pure radiance of the past.'

Petrarch, the 14th-century Italian poet, looks back to the glories of the classical world. It was he who coined the misleading term 'Dark Ages' for the centuries between the fall of Rome and what became known as the Renaissance

The positioning of humans rather than God at centre stage was echoed in the visual arts. Now wealthy secular patrons wanted architects to build them palaces, and to fill them with sculptures and paintings based on classical mythology rather than scenes from the Bible. However, some of the most magnificent Renaissance art was still religious. The Roman Catholic Church remained one of the biggest patrons of the arts – as seen at St Peter's in Rome and Michelangelo's frescoes for the Sistine Chapel. The Church was enormously rich, owning vast tracts of land from which it drew extensive revenues. It also made money by selling pardons to sinners, a practice called simony.

Some Christians felt the Pope and the vast Church hierarchy had grown far too worldly, and called for a return to the simplicity of the early Church. In particular, they denounced the practice of simony. In 1517 the German monk Martin Luther nailed an attack on simony to the door of the castle church at Wittenberg, a key step in the first successful revolt against Church authority – the Reformation.

It was not just corruption that Luther and his followers attacked. They believed that the claim of priests and the whole Church hierarchy to act as mediators between the individual and God was wrong. Until now the Bible had been available only in Latin, and the Church claimed the sole authority to interpret it to the people. Lutherans insisted that the Bible should be translated into vernacular languages, so that everybody could know and interpret the word of God. They denied the priesthood any special status, believing that each individual should stand on their own, face to face with God.

The Church moved to suppress the Protestant reformers, but these had their supporters among the princes of Europe, some of them ambitious to gain more control over their local Church and its wealth. Religion thus became mixed up in power politics. Long drawn-out wars followed, as Europe descended into more than a century of bloodshed.

THE LONG ROAD TO TOLERATION

Religious intolerance is almost as old as religion itself. Where belief is couched in terms of a battle of good versus evil, it may entail a sense of righteous certainty, a conviction that anyone who holds different beliefs is damnable and deserves death.

The persecution of dissenters is most likely to happen where religion has been institutionalized. Whenever this happens, persecution preserves power as well as defending doctrine. In many cases, institutionalized religion has been allied with (or even controlled by) state power. One title of the Roman emperors was *pontifex maximus*, 'high priest'. The Romans tolerated a range of religions, many even joining such cults as Mithraism. But the Roman state saw the early Christians as a threat: they attracted the dispossessed, and talked seditiously about building Christ's kingdom on Earth. Persecutions ensued, until the Roman emperors decided to adopt Christianity themselves. Now the Church became an instrument of state power, and in its turn persecuted religious minorities regarded as heretical.

Sometimes those in power have recognized that toleration, original thought, innovation and pluralism make for a happier, more prosperous society. In imperial China, three religions – Confucianism, Daoism and Buddhism – coexisted peacefully. In India in the later 16th century, Akbar, the greatest of the Mughal emperors, sought to hold together his vast empire by extending religious toleration to all his subjects, whether Hindu, Sikh or Jain, even though he himself was a Muslim. But a century later his powerful descendant Aurangzeb turned the Mughal empire into a more exclusively Muslim state, repressing the Hindu majority and fighting wars against the Sikhs, whose ninth Guru he had executed. He thus damaged the sense of imperial unity built up by Akbar, and in the century after his death in 1707 Mughal power was fatally eroded, allowing Europeans to gain a foothold on the subcontinent.

'The emperor's court became the home of the inquirers of the seven climes, and the assemblage of the wise of every religion and sect.'

Abul Fazl, *The History of Akbar* (c. 1590), pointing out the benefits of the Mughal emperor's religious toleration

Muslim rulers in Spain had at first shown tolerance towards Jews and Christians, permitting them to practise their religions as long as they were prepared to pay higher taxes. The caliphate of Córdoba (929–1031) witnessed something of a golden age, as culture flowered and trade expanded. But toleration shrank under the Almohad dynasty in the 12th–13th centuries, and vanished after the Spanish Christians completed their *Reconquista* of the Iberian peninsula by seizing Granada in 1492. Jews and Muslims were forced to convert to Christianity or face expulsion, and this drove many of the most intelligent and skilled inhabitants abroad – just as when Louis XIV turned against French Protestants in 1685, and when the Nazis persecuted German Jews in the 1930s.

The Reformation launched in the early 16th century set Protestants against Catholics, and over the next two centuries Europe suffered a frenzy of persecution and conflict. The Thirty Years' War (1618–48) killed as much as one-third of the German population, mostly through starvation and disease, as the armies of the Catholic and Protestant powers of Europe laid waste to the land. It took many generations for Germany to recover.

Religious toleration only became the norm much later across Europe, despite the urgings of the thinkers of the 18th-century Enlightenment (see p. 162). In Britain, not until the Crown Act of 2013 could a Roman Catholic take the throne. To this day religious toleration faces threats in many nations, both from fundamentalists and from authoritarian secular states such as China.

PRINTING

If the first communications revolution came with the development of writing, the second arrived with printing. When every text had to be copied by hand, only a very few manuscripts could circulate. The invention of printing with movable type brought a huge proliferation in the number of different texts published, and a much wider distribution of them. The social, cultural and intellectual impact was enormous.

The Chinese had started to use woodblocks to print texts and decorative designs on textiles and paper (their own invention) in the 3rd century CE. On the woodblocks the wood was carved away to leave characters or images protruding in relief. By the 9th century whole books were being printed, and by the 14th century individual blocks had been carved for each of the 80,000 Chinese characters, which could then be recombined to make a page. This is the principle of movable type, but, with so many characters, the Chinese stuck largely with woodblock printing, each page carved as a unit. In Korea, printing with movable metal type seems to have been introduced in the 14th century. Literacy rates in East Asia were higher than in the rest of the world until about the mid-19th century.

'If we think to regulate printing, thereby to rectify manners, we must regulate all recreations and pastimes, all that is delightful to man.'

John Milton, *Areopagitica* (1644)

The simplicity of the Roman alphabet, with its limited number of characters, is well suited to the principle of movable type, and in the mid-15th century the German goldsmith and publisher Johannes Gutenberg introduced a system that stayed in use for 500 years. Rather than assembling pages out of individually carved woodblock letters, he used type made out of an alloy of lead, tin and antinomy, with a low melting point. Once he had made a mould for a character or punctuation he could cast all the copies he liked. These he arranged side by side along a strip of wood to make up words in lines, and the lines were justified (spread out to a uniform width) by inserting wedges of metal between the words. It took about a day to set up a page of type, then a vice-like press transferred ink from the printing plate to the paper. In 1455 Gutenberg printed his first book, the Latin Bible. The new technology spread rapidly, and in 1475 William Caxton printed the first book in English. By the end of the century several million books had been printed in Europe. By 1800 this figure had risen to 2 billion copies.

The ensuing torrent of ideas, knowledge and opinions via books, broadsheets, ballads and pamphlets gave rise to a surge in literacy – till then the province of churchmen and a small secular elite. Printing meant that the classical texts rediscovered by Renaissance humanist scholars would no longer remain in darkness. It also allowed the ideas of the Protestant reformers to spread like wildfire through Europe, widening the support base of the Reformation.

Small wonder that those who held power were extremely suspicious of this democratization of knowledge, and the opportunities it created for criticism and political activism. Most states – and the Roman Catholic Church – attempted to dictate what could and could not be published and read. But even where books were censored or burned, the technology of printing meant that someone somewhere would produce more copies.

THE SCIENTIFIC REVOLUTION

During Europe's so-called 'Dark Ages', the learning of the ancient Greeks was mostly forgotten, or condemned as pagan. Much of their science survived only through Arab scholars, who also made significant contributions to subjects such as mathematics and chemistry (the words algebra and alcohol derive from Arabic).

On the other side of the world, China was a hotbed of technological innovation, whose range of inventions included the magnetic compass, gunpowder, papermaking and printing. Eventually these 'Four Great Inventions' reached the West.

Even when ancient Greek learning was rediscovered in Europe, it didn't immediately inspire new thinking. Scholars regarded the ancient Greeks as the ultimate authority, especially once theologians had written their version of Greek thought into Roman Catholic doctrine. To question this authority was heresy.

A central tenet of Church teaching was that the Earth, on which God had created man, lay at the centre of the universe. This echoed the cosmology of the Greek geographer Ptolemy (1st century CE), although an earlier scientist, Aristarchus of Samos (3rd century BCE), had proposed that the Earth orbits the Sun. This heliocentric theory was revived in the 16th century by the Polish astronomer Nicolaus Copernicus. Although both mathematics and observations confirmed it, he did not dare publish his findings till 1543, the year of his death. When the Italian physicist and astronomer Galileo Galilei produced evidence that backed Copernicus, the Roman Catholic Church put him on trial, and in 1633, under threat of being burned as a heretic, he withdrew his support. Nevertheless, Galileo's contributions to modern science were enormous, particularly his use of mathematics in physics.

> **'I am much occupied with the investigation of physical causes. My aim in this is to show that the celestial machine is not similar to a divine animated being, but similar to a clock.'**
>
> Johannes Kepler, letter to his patron (1605). Kepler built on the findings of Copernicus, and found the laws of planetary motion

Combining observation and experiment with mathematical analysis became the hallmarks of the new scientific method. General theories were to be derived from particular observations of the real world – a method triumphantly vindicated when in 1687 Isaac Newton published his law of gravity and three laws of motion, which described the interactions between forces and objects. The stress on a mechanized cosmos identified regular and predictable processes that could be mathematically defined. The prestige of Newtonian ideas helped ensure that the concepts, methods, language and metaphors used to explain them were applied in various branches of knowledge.

Breakthroughs took place in other fields of science. In the year when Copernicus published his heliocentric theory, the Flemish anatomist Andreas Vesalius published *On the Workings of the Human Body*, based on dissections he had himself performed rather than on the teaching of the ancient Greek physician Galen – till now the ultimate authority in such matters. Galen's medical theories were also challenged by the 16th-century Swiss-German physician Paracelsus, who began the move from medieval alchemy towards modern chemistry, and insisted that specific diseases require specific remedies.

Galen had followed the principle of Aristotle that the world is made up of a balance of four elements (earth, water, air and fire). Newton's contemporary, Robert Boyle, promoted the altogether different concept of chemical elements. Both Boyle and Newton belonged to the Royal Society of London, founded in 1600. It was just one of many academies of science that were established around Europe in the 17th and 18th centuries. Not so many years before science had flown in the face of authority. Now it became a respectable activity for a gentleman.

EUROPE EXPANDS

'Printing, gunpowder and the mariner's needle . . . have changed the whole face and state of things throughout the world.' So wrote the English philosopher Francis Bacon in 1620, looking back over a century and a half in which the horizons of Europe had broadened in ways that would have previously been unimaginable.

Ironically, the three inventions Bacon lists all originated in China, and the process he describes was largely a matter of Europe beginning to catch up with the world's richest and most technologically sophisticated superpower. Although the Chinese had used gunpowder for fireworks for centuries, and had started making guns, it was the Europeans who seized on this new technology, making the first cannon in the 14th century. The Chinese had long possessed the 'mariner's needle' (the magnetic compass), and for centuries had traded all around the Indian Ocean. The compass was introduced to the Mediterranean by the Arabs, and adopted by European mariners, who developed other navigational instruments such as the quadrant and the astrolabe.

In the early 15th century the Chinese admiral Zheng He embarked on a series of state-sponsored expeditions to Indonesia, India, Arabia and East Africa, but in 1433 Chinese imperial policy went into reverse, and the voyages stopped. This was partly down to costs and to factional politics, but the Chinese mandarins (senior civil servants) also seem to have decided that China was rich enough in natural resources, and thus the emperor need not demean himself by engaging in overseas trade.

Western princes and merchants had other ideas. With the closure of the overland Silk Road to the Far East by the Ottoman Turks (see p. 102), those European countries on the Atlantic seaboard seized their chance to seek maritime routes round Africa to the East Indies, whose spices were precious commodities in the medieval and early modern world. In the mid-15th century Prince Henry of Portugal set up a school of navigation, and sponsored voyages down the west coast of Africa. Portuguese exploration continued after Henry's death in 1460, and in 1488 Bartolomeu Dias rounded the Cape of Good Hope and entered the Indian Ocean. Ten years later another Portuguese sailor, Vasco da Gama, reached India by this route. The Portuguese went on to set up trading posts all round the African coast and across southern and eastern Asia, as far as China and Japan.

Even more ambitiously, sponsored by the king and queen of Spain, in 1492 Christopher Columbus sailed west across the Atlantic, believing this would prove a shorter route to the East Indies. When he found the islands of the Caribbean, he believed he had reached his goal (which is

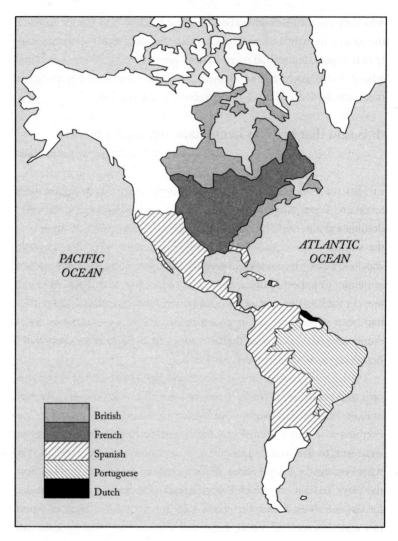

	British
	French
	Spanish
	Portuguese
	Dutch

European colonization of the Americas, c.1750

why they are misleadingly called the West Indies). He went on to reach the American mainland, and realised that this New World was immensely rich in gold. It also had many people whose souls could be saved and whose bodies could be put to work. And with God and gunpowder on their side, how could the European conquistadores fail?

'It is said that hell has been made manifest on earth.'

Japanese account of Portuguese slave trading (1580s)

In 1494 the Pope brokered a treaty dividing the New World between Spain and Portugal: Portugal gained Brazil, and Spain the rest. Europe's domination and exploitation of the Americas – and much of the rest of the world – was about to begin. The brutal conquest of the Americas was followed by the death of much of the native population who lacked immunity to imported diseases such as measles and smallpox. Within a few decades 90 per cent of a population of between 50 and 100 million had been wiped out. The conquerors had guns, horses, armour and a dominant faith, but it was their microbes that did most to destroy the peoples and cultures of the New World.

The need for a viable labour force encouraged the Europeans to develop a transatlantic slave trade. In the most squalid conditions they shipped millions of captives bought from local traders and rulers in Africa. The Portuguese colony of Brazil was the largest destination but many slaves were sent to the West Indies and a smaller number were consigned to what became the United States. Many European merchants profited from the 'Slave Triangle', by which European manufactured goods were traded for slaves in West Africa, the slaves sold in the New World, from where raw materials such as cotton, sugar and tobacco were sent back for sale in Europe (see map, p. 101).

The European empires in the Americas and the creation of a maritime route to southern and eastern Asia gradually transformed much of the world. For example, although most of the world's major cities in the 16th century lay in Asia, European maritime cities such as Lisbon and Seville grew to global stature.

The process continued in the 17th century. In Europe, Amsterdam and London grew in stature as the centres of empire. The Dutch and British also founded new cities overseas – Quebec in 1608, New Amsterdam (later New York) in 1614, Cape Town in 1652.

THE ENLIGHTENMENT

In the 18th century, many thinkers across Europe began to question the dogmas and authority of state-sanctioned religion and to hold up the virtue of reason as an alternative to superstition. 'Superstition sets the whole world in flames, philosophy quenches them,' wrote the French writer Voltaire, one of the leading figures of the loose intellectual movement called the Enlightenment.

The achievements of scientists such as Isaac Newton and the writings of the English philosopher and political theorist John Locke (see pp. 157 and 169) were both influential. Newton and Locke embraced empiricism, the belief that knowledge derives from observation and experience, not from innate ideas. The former is the realm of reason, Locke held, while the latter is the realm of faith. Everything is up for question, all assumptions must be critically examined.

The thinkers of the Enlightenment sought to establish a rational basis for all human knowledge and action, from economics and the law to psychology, education and history. Although few were out-and-out atheists, many subscribed to deism, which restricts the role of God to the 'first cause' of the universe. In France these thinkers were known as the *philosophes*, and included such figures as Voltaire, Montesquieu and Denis Diderot (editor of the innovative *Encyclopédie*, 'a reasoned dictionary of the sciences, arts and crafts'). A similar movement arose in Britain, particularly in Scotland, with figures such as the sceptical philosopher David Hume and the economist Adam Smith (see p. 171). All these figures are notable both for upholding the value of tolerance, and for their humanitarianism.

It took time for the values of the Enlightenment to be more widely adopted. Various European monarchs paid lip-service to the Enlightenment in the 18th century, but continued to rule as autocrats. The US Declaration of Independence of 1776 and the Bill of Rights of 1789 enshrine certain values of the Enlightenment, as does the Declaration of the Rights of Man and of the Citizen adopted in the wake of the French Revolution of 1789. However, the Enlightenment ideals of reason and tolerance still have not taken hold in many parts of the world.

The fate of free-thinkers

'*Écrasez l'infâme*,' wrote Voltaire – 'Crush superstition.' He had in mind the treatment of those who would have considered themselves free-thinkers, but whom Church and state condemned as blasphemers. In 1697 an Edinburgh student called Thomas Aikenhead became the last person in Britain to be executed for blasphemy – he had called theology 'a rhapsody of ill-invented nonsense'. In 2012, thirty-three countries around the world still had anti-blasphemy laws, and in a number of Muslim countries the penalty for blasphemy is death.

'There is no effectual way of improving the institutions of any people but by enlightening their understandings.'

William Godwin, *An Enquiry Concerning Political Justice* (1793)

THE INDUSTRIAL REVOLUTION

The term 'Industrial Revolution' describes a gradual process of change that took place, chiefly in Europe and North America, in the 18th and 19th centuries, and then spread to much of the rest of the world.

Manufacturing of one kind or another had existed for millennia, of course, from the hand-axe industries of the Neolithic period to the bricks of Babylon. By the 18th century just two regions – the Indian subcontinent and China – were responsible for as much as three-quarters of the world's manufacturing output. High-quality textiles and porcelain were exported as far as Europe. Most of this manufacturing took place in small-scale workshops, frequently rural rather than urban. Europeans had done much the same for many centuries. The workshop was only gradually replaced by mechanization and the factory system – the key features of the Industrial Revolution, together with mass urbanization. What followed was soaring productivity, and a huge surge in population due partly to a lowering of the age at which people had children, thanks to the independence conferred by wage labour.

Why did the Industrial Revolution begin in Britain? The country was politically united and stable, free from internal customs barriers (unlike much of Europe), and had an advanced banking system. Britain also profited from its geographical position on the Atlantic seaboard and its aggressive use of its navy to become, during the 18th century, the world's leading mercantile power, far outstripping the Asian coastal economies. The trade in goods such as cotton, tobacco, sugar and slaves made huge profits for many merchants, who put away capital to invest in new industrial enterprises. Britain's many ports and its many navigable rivers (backed later by a network of canals), fostered both external and internal trade.

Britain's own natural resources also played a key part, especially its reserves of iron and coal. Previously iron had been smelted using charcoal, but the change to coke (derived from coal) brought a massive rise in output. Coal was also key to the technology that really got the Industrial Revolution going: steam. China and India, the other major centres of early industry, could not make comparable use of coal.

'I sell here, sir, what all the world desires to have – Power.'

Matthew Boulton, partner in James Watt's steam engine business,
as recorded by James Boswell (22 March 1776)

All kinds of ingenious new machines for spinning and weaving were developed during the 18th century. They needed less skilled operators than the older manual methods, and could turn out textiles in greater and greater quantities. At first the machines were driven by water power, particularly on the Pennine slopes of northern England, but coal-fired steam power enabled the construction of factories in many more locations. Steam engines had been used to pump water out of mines since the 18th century, but James Watt's improvements in the last quarter of that century spread the technology to factories, and later to ships and locomotives, allowing the mass transit of both people and goods (see p. 180).

Other countries in Europe followed fast, notably France, Belgium and Germany. By the start of the 20th century Japan and Russia were also significant players, but the true industrial superpower – already the largest by the late 19th century and ready to dominate the 20th – was the USA. It benefited from plentiful natural resources, particularly coal, and from an entrepreneurial culture and society.

Although the Industrial Revolution was to bring unprecedented economic growth and eventually a real increase in wages, it inflicted considerable social costs. Working conditions were all too often dangerous and alienating, and living conditions all too often plunged the depths of squalor (see p. 179).

THE AGRICULTURAL
REVOLUTION

The Industrial Revolution would not have been possible without the Agricultural Revolution. Both were gradual events, and took place across the world at different times and speeds. But without the increased food yields made possible by agricultural improvements, societies could not have been transformed from rural agrarianism to urban industrialization.

Like the Industrial Revolution, the Agricultural Revolution started in Britain but then spread elsewhere. Between 1650 and 1800, British agricultural productivity almost doubled. One factor was the amount of land under cultivation, which grew by around 20 per cent during the 18th century, partly because of a process called 'enclosure', by which wealthier farmers and landowners took over common land or open fields to which peasant farmers could prove no legal title. By the mid-18th century much of England's farmland had already been enclosed. The peasants either had to become waged agricultural labourers, or seek work in the cities. Similar changes occurred elsewhere in Europe (for example in Prussia). Even where peasant farming persisted (as in France), there was a shift from subsistence to market production.

Although such changes damaged the peasantry, they improved food production. Wealthy landowners carried out improvements on the land over which they now had sole control. They drained marshes and erected field boundaries, enhancing the selective breeding of livestock. With new rotation systems the fields could be used every year, rather than left fallow to recover their nutrients. Fodder crops like turnips were grown, and kept many more animals alive through the winter. In the old days most had been slaughtered, and their meat salted.

'But a bold peasantry, their country's pride, When once destroyed, can never be supplied.'

Oliver Goldsmith, *The Deserted Village* (1770)

Technology revolutionized agricultural output. The Oxfordshire farmer Jethro Tull introduced the seed drill in 1700. In Prussia in 1747 the first sugar was extracted from beet. Cane sugar had been a luxury imported from the West Indies; now it was to become part of the staple Western diet. In 1785 the cast-iron ploughshare was patented in Britain, followed in 1800 by power-driven threshing machines, and in 1830 by reaping machines. In the 19th century Chile exported bulk quantities of guano, fertilizer extracted from accumulations of seabird droppings.

As the population increased, and the proportion working on the land decreased, Europe needed to import more food. So began a drive to turn other parts of the world to food production. John Deere's development of a steel plough in 1837 enabled the cultivation of the hard soils of the North American prairies, which began to export huge quantities of wheat – although at the expense of the native Americans, who were moved off their traditional hunting grounds into 'reservations'. The introduction of the petrol-driven tractor in 1892 further boosted productivity. By the end of the century in the USA, it took less than one-third of the man-hours to produce a tonne of wheat than in 1800.

With the coming of railway networks, faster steamships, and canning and refrigeration technologies, the later 19th century introduced large-scale livestock farming into North and South America, Australia and New

Zealand, all of which exported to Europe, where meat could feature more often in everyday diets.

This is not to say that nutrition for many working-class people was remotely adequate. In many places the staple diet consisted of potatoes or bread (perhaps with some dripping). Even though starvation was largely a thing of the past in Europe by 1900, malnutrition was widespread. In 1899 three out of every five volunteers for the British army – drawn largely from the unskilled working class – were rejected as medically unfit. In many other parts of the world, peasants still depended on a single staple crop. In China and India, for example, the majority had to subsist – as they had done for centuries – on a single bowl of rice per day. And famine remained an ever-present threat.

THE SOCIAL CONTRACT

The idea that rulers rule by divine sanction, rather than by consent of the ruled, goes back to the origins of kingship. Some of the earliest texts of the ancient Near East are genealogies that trace the ruler's ancestry back to a god, so justifying their earthly power.

In Egypt, the pharaoh was the son of the Sun god Ra. In Japan, emperors claimed descent from the goddess Amaterasu, a claim abandoned only after the country's defeat in the Second World War. In China the emperor held 'the mandate of heaven', but an unjust emperor might find its mandate removed. This concept fanned the many violent changes of dynasty in Chinese imperial history.

In the Judaeo-Christian tradition, the monarch is anointed with oil at the coronation. This derives from the biblical account of the anointment of David, king of Israel: 'Then Samuel took the horn of oil, and anointed him in the midst of his brethren: and the Spirit of the Lord came upon

David from that day forward.' The idea of the monarch as God's anointed led to the doctrine of 'the divine right of kings', which holds that the king needs no consent from his people, his aristocracy, his parliament, or even the Church. He answers only to God.

King James VI of Scotland (later James I of England) was one monarch who held to this absolutist dogma. His beliefs led him to disregard the rights and privileges claimed by the English parliament – the legislative assembly intended to represent the people. His son and successor, Charles I, shared his views, and tried to rule without parliament altogether. The upshot was civil war and the execution of the king in 1649 for treason against his own people.

It was the chaos and bloodshed of the English Civil War that led Thomas Hobbes to publish his book *Leviathan* in 1651, proposing the idea of a social contract between ruler and ruled. For human beings in a 'state of nature', Hobbes argued, life had been 'solitary, poor, nasty, brutish, and short'. To avoid such barbarism, humans had come together and agreed to a social contract by which, in return for protection, they surrendered some rights to an absolute authority.

What this implied was that if the absolute authority failed on its side of the bargain, people had the right to replace it. This was spelled out in the version of the social contract proposed by another English philosopher, John Locke, in his *Treatises of Government* (1690). He argued that government is legitimate only where it has the consent of the governed. The state guarantees to preserve the 'natural rights' of the citizens, specifically life, liberty and property. Should the government break this social contract, then the people may choose another ruler – an argument deployed by the American revolutionaries of 1776, who decided to replace a British king with an independent republic.

A third version of the social contract came from the French philosopher Jean-Jacques Rousseau in his book *Du Contrat Social* (1762). Rousseau argued against the principle of representative government found in Britain's constitutional monarchy, and stated that liberty could only exist where the people as a whole were directly involved in making the laws, which should express the 'general will'. In a small state this might be achieved through direct democracy, but in larger states, Rousseau argued that the general will required the guidance of a strong government. However since the government would always seek to strengthen itself, the people should periodically be able to alter the form of government and replace its leaders.

Rousseau's ideas inadvertently encouraged some of the more tyrannical leanings of French revolutionaries such as Maximilien Robespierre, who in 1792 declared 'I myself am the people' – a mantra echoed by some of the most murderous dictators of the 20th century.

The US Declaration of Independence

The famous preamble of this historic document, drafted by Thomas Jefferson, makes specific use of Locke's concept of the social contract: 'We hold these truths to be self-evident, that all men are created equal; that they are endowed by their Creator with inherent and inalienable rights; that among these, are life, liberty, and the pursuit of happiness; that to secure these rights, governments are instituted among men, deriving their just powers from the consent of the governed; that whenever any form of government becomes destructive of these ends, it is the right of the people to alter or abolish it, and to institute new government . . .'

FROM MERCANTILISM TO FREE-MARKET CAPITALISM

Those European powers that began to build overseas empires from the 16th century on were aiming to maximize their share of a growing and profitable sphere: international trade. According to the then-established theory of mercantilism, the amount of wealth in the world was fixed. As a consequence, each of the leading European powers aimed to grab the largest possible share of international trade.

This stoked a series of wars around the globe. The Seven Years' War of 1756–63, for example, saw fighting between European powers, especially Britain and France, not only in Europe, but as far afield as the Caribbean, North America and India. These powers also passed protectionist measures that barred all but the homeland's citizens from participating in trade to and from the mother country. Not even their own overseas colonists were allowed to benefit from this trade. According to mercantilism, colonies were established solely for the benefit and profit of the mother country. Such policies eventually drove Britain's colonies on the eastern seaboard of North America to declare their independence as the United States in 1776.

The same year saw the publication in Britain of *The Wealth of Nations*, by the Scottish philosopher and political economist Adam Smith. This book is seen as the founding text of free-market capitalism. Up to this point, international trade had been subject to such constraints as prohibitive excise duties or naval action against foreign merchant ships. In addition, commerce within a particular country was often inhibited by state control. This might operate through taxes, or even internal customs duties, but it often took the form of royal patents: in return for a sizable payment, the monarch would grant the sole right to provide a particular good or service. No other person was entitled to infringe this monopoly, and the holders could charge whatever price they wanted.

Adam Smith held that all these restrictions on the freedom of the market were inefficient. He argued that if individuals are allowed to pursue their own economic self-interest, then the laws of supply and demand, which he called 'an invisible hand', would augment not only the wealth of nations but also the prosperity and thus the happiness of the citizens of those nations. But the laws of supply and demand, he argued, could only work successfully in a free market. And such a market should work not only within nations, but between them.

Free-market capitalism became the norm within many industrialized countries in the 19th century. However, it became apparent that companies that outcompeted their rivals would tend to become monopolies, which would allow them to dictate the price of their products. Even in the USA, that champion of free-market capitalism, the government felt obliged, from the 1890s, to rein in the unbridled free market by passing anti-trust legislation, so as to break up large companies that threatened to monopolize certain markets. And in the 20th century many developed countries chose to further regulate their industries by imposing health and safety standards on employers.

'Every individual necessarily labours to render the annual revenue of society as great as he can. He generally neither intends to promote the public interest, nor knows how much he is promoting it. He intends only his own gain, and he is in this, as in many other cases, led by an invisible hand to promote an end which was no part of his intention.'

Adam Smith, *The Wealth of Nations* (1776)

Following Adam Smith, in the 19th century economists increasingly advocated free trade between nations, unhindered by such things as import duties. However, agricultural producers and the industrial manufacturers often argued in favour of such duties, so that they would be protected from foreign competition. As a consequence, protectionist measures continued to restrict international trade well into the 20th century. Even the establishment of free-trade areas such as the European Common Market only benefited members.

Since the later 20th century there have been concerted international efforts to break down protectionist barriers and create a genuinely globalized market. Some argue that this creates a new imbalance of power: large multinational corporations can dominate the market to the detriment of smaller companies, workers and consumers worldwide. While free trade and free markets have helped to foster global economic growth, the problem of setting the ideal balance between free markets, protectionism, market regulation, consumers' interests and employees' welfare is still a fundamental political and economic debate today.

NATIONALISM AND THE NATION

From the 18th to the 20th centuries many societies saw radical transformations in the form of industrialization, urbanization and rising rates of literacy. These brought a greater awareness of politics, a remoulding of political values, and the rise of new ideologies.

One of the most important of these new ideologies was nationalism. Nationalism – or at least a belief that the good of a given nation mattered most – had long been present in powerful sovereign states such as Russia and China, which owed no allegiance beyond their own frontiers.

The Chinese had long referred to their empire as 'the Middle Kingdom', implying that it was the centre of the world. In Europe, the rejection of papal authority by many Protestant princes during the Reformation reflected a desire for complete control within a state's own borders. But such versions of nationalism tended to represent only the interests of the ruling élites. Long-established European empires such as that of the Habsburg family, which ruled over a wide mix of nationalities, defined themselves largely in dynastic terms.

France and England had for some centuries sought to define themselves as nation-states, but again this came largely from the centre. Although within the frontiers of France people spoke many different languages, from Breton to Basque, kings and governments long upheld the sole use of French, and in 1635 Louis XIII established the Académie française as the 'guardian' of the language. England, occupying as it did the greater part of the island of Britain, separate from the European mainland, had revelled in its isolation since the time of Shakespeare, who famously celebrated 'this sceptred isle', and whose language, together with that of the Authorized Version of the Bible (1611), commissioned by King James I, did much to forge an English national identity.

In 19th-century Europe, pressures for change gave rise to growing national demands in European countries that had long been part of larger empires – for example in Hungary, Ireland and Poland, ruled respectively as part of the Austrian, British and Russian empires. There was a large-scale, but ultimately unsuccessful, rising in Hungary against Habsburg (Austrian) rule in 1848–9, and risings in Poland against Russian rule in 1830 and 1863. In Ireland, the failed rebellion in 1798 against British rule (which had begun in the 12th century with seizures of power and land by Anglo-Norman adventurers) was followed in the 19th and 20th centuries both by large-scale popular agitation and by smaller-scale armed insurrection.

In the earlier years of the 19th century, many strands of nationalism were rooted in ideals of liberty and equality, and aimed to locate political legitimacy within nations that were 'natural' and cohesive. These beliefs were connected to the constitutionalist idea that laws should restrict the power of government, because its legitimacy derived from 'the people'.

So nationalism was more than a single struggle for new territorial identities and boundaries. It also involved efforts to define what a nation really was, and whether it was rooted in ethnic, linguistic, geographical or other common factors. Many nationalists thought in cultural terms, and encouraged the use of indigenous languages. Intellectuals sought to identify the inherent features of nations, and so of national communities. These trends tended to ignore the porous nature of both natural and political borders across Europe. Ethnic and linguistic groups were not always located within specific territories. German-speaking communities were strewn across Central and Eastern Europe (already a great melting pot of peoples), including many parts of Russia. France might present itself as a monocultural nation-state, but it included many different ethnic and linguistic groups. And the United Kingdom of Great Britain and Ireland, created in 1801, contained speakers of various Celtic languages as well as English.

Linked to this quest for 'the people', many poets, composers and other artists based works on distinctly national or ethnic forms. For example, the composer Antonín Dvořák (1841–1904), like other Czechs a subject of the Habsburgs' Austro-Hungarian empire, aimed to integrate elements of traditional Czech folk music into his work. All across Europe there was a growing interest in folklore and 'national' language. In early 19th-century Germany, stunned by recent defeats at the hands of Napoleon's armies, the Brothers Grimm collected 'authentic' German folk tales and worked on a definitive dictionary of the German language.

Both Germany and Italy had long been patchworks of smaller states, with both foreign and native rulers. Although artists and intellectuals, liberals and democrats, had for decades called for national unification, this largely came about as a result of armed conflict in the mid-19th century. These campaigns were followed by pressure in the Balkans to throw off Turkish rule, pressure that had already led to independence for Greece in 1830. Serbia, Romania and Bulgaria followed. The fashionable nature in western Europe of some causes – Greek independence in the 1820s, the Italian *Risorgimento* ('resurgence') from the 1840s – showed that specific nationalisms could win international support when other states did not feel threatened.

'Not through speeches and majority decisions will the great questions of the day be decided . . . but by iron and blood.'

Otto von Bismarck (1862). Bismarck, then chief minister of Prussia, was referring to earlier, failed efforts by liberals and democrats to create a united German nation. Over the following decade he was to mastermind the reunification of Germany by military force.

The First World War, which brought down the Austrian, German and Russian empires, created many new European nation-states, among them Poland, Czechoslovakia, Finland, Estonia, Latvia and Lithuania. The idea that state boundaries should follow those of national groups developed as a new norm, at first in Europe, and then in the rest of the world. As a yearning for national self-determination spread to European colonies such as India, imperial structures and ideals gave way. By 1975 most of Europe's overseas empires had ceased to exist. In many regions, particularly in Africa and the Middle East, the frontiers inherited from the colonial carve-up by newly independent states proved to be arbitrary, frequently ignoring local or tribal identities or boundaries. This has often subsequently led to civil war and ethnic conflict.

Nationalism may seek to be inclusive, uniting a particular 'people', but it is also inevitably exclusive. If you do not share a nation's self-identity, you do not belong, and this has bred discrimination, if not more violent action, against members of ethnic and religious minorities. For example, the Turkish nationalism that became the dominant force within the Ottoman empire in the early 20th century had severe and sometimes lethal outcomes for its Armenian, Greek and Kurdish subjects: not being ethnic Turks classed them as aliens. This treatment contrasted with the more inclusive view that had formerly cemented the polyglot Ottoman empire.

Nationalism could also be linked to economic protectionism and to its cultural equivalents. For example, the pressure to produce works in national or would-be national languages often arose as a gesture of defiance against foreign rulers. But in the longer term it could cause crippling cultural isolationism.

National self-determination

On 8 January 1918, in the last year of the First World War, President Woodrow Wilson set out the war aims of the USA. His 'Fourteen Points' listed the basic principles on which he wanted the war to be resolved and disputes to be settled thereafter. On 11 February he went on to say: 'National aspirations must be respected; people may now be dominated and governed only by their own consent. "Self-determination" is not a mere phrase; it is an imperative principle of actions . . .'

Although nationalism fuelled the rhetoric of colonial peoples struggling to be free from European rule, its consequences in the 20th century were often pernicious. Most notoriously, in Germany the National Socialists (Nazis) used it to justify tyranny, war and genocide. In more recent times, nationalism has continued to ignite ethnic violence – as it did during the creation of new nation-states in the former Yugoslavia in the 1990s. 'Ethnic cleansing' became the new euphemism for the murder or expulsion of unwanted minorities, and the refugee became one of the characteristic figures of the modern world.

URBANIZATION

In the ancient world, cities such as Nineveh, Babylon and Alexandria had reached populations of 100,000, and Rome was probably the first to reach a million. While many cities in later centuries matched the size of Rome, the 19th century saw a spurt in urban growth. By 1900 both London and New York surpassed 5 million. Many other world cities grew at similar rates. A key factor was industrialization, which drew vast numbers from country to city.

There was at the same time a big rise in global trade, especially across the Atlantic, with both primary products such as grain and manufactured goods exported in bulk to Europe from the Americas, so that coastal cities such as New York and Buenos Aires grew significantly. There was comparable development – though later and at first on smaller scales – in the Pacific world for cities such as San Francisco, Sydney, Singapore, Tokyo and Hong Kong.

Being at the hub of railway networks also encouraged the growth of inland cities such as Chicago. Railways also led to suburbanization, which altered the shape of cities and towns by enabling many of the newly wealthy middle classes to move into leafy outlying suburbs.

The many new factories, usually located in towns and cities, required substantial workforces. Capital cities such as Berlin also grew as government itself became an employer of more and more people.

Cities created an increasingly manmade environment, no longer ruled by the rhythms of rural life. This was vividly demonstrated when street lights challenged the dark and sewers replaced the collection of human excrement by cart for use as fertilizer. The symbolic power of cities was expressed when they hosted great displays of technology and power, beginning with London's Great Exhibition of 1851.

Cities also posed serious problems for health and living standards. Slums, poor sanitation and overcrowding encouraged diseases such as cholera and tuberculosis. Cities were also hotbeds of popular uprisings against the ruling classes, as happened in several European states in 1848. Many urban 'improvement' schemes were designed to destroy slums where the authorities felt they had lost control. It has been argued that Haussmann's redesigned boulevards in Paris, influential in cities worldwide, were an attempt to create an urban environment that favoured the army over the mob: wider streets were more exposed and harder to barricade.

Under Western rule or influence, many non-Western cities also acquired features such as railway stations, boulevards, telegraph buildings and major hotels. Nevertheless, in 1900 the largest urban areas were still concentrated in Europe and North America.

'Hell is a city much like London – A populous and smoky city.'

Percy Bysshe Shelley (1819)

EXPANDING HORIZONS

The 19th century saw expanding horizons in thought, expression and experience. Progress in science and technology – from the harnessing of electrical power to the creation of new synthetic materials – enhanced the belief that human life on Earth could steadily improve. At the same time, faster modes of transport made the world a smaller, more interconnected place.

The steam locomotive enabled humans to travel at speeds once unimagined, and allowed for the long-distance mass transportation of both goods and people – as did the subsequent development of the steamship.

The invention of the telegraph, and subsequently the telephone and wireless, made long-distance communication virtually instantaneous. Such achievements led people to imagine a future hugely different from either past or present. Anything seemed possible. When in 1903 the first flight in a heavier-than-air aircraft was made, it was clear that not even the sky was the limit.

Access to this expanded world varied greatly. Many in the 19th century still lived in villages and followed the lifestyle of their ancestors. And yet across societies, even among the poor, more and more people moved from the place where they were born.

The expansion of global steamship lines was instrumental in a massive growth in migration. Europeans migrated to the Americas and to Australasia, while Chinese crossed the Pacific, especially to California, and Indians moved to work in South Africa, Fiji, Trinidad and other far-flung places. These movements combined to transform the world's population patterns.

The expansion of imperial systems and of transport links boosted the trade in animals and plants grown for human consumption. Rubber, first

The Wright Brothers' first flight, 1903, North Carolina, USA. Wilbur Wright is lying on the plane's wing, while his brother Orville is running at the wingtip

harvested from wild trees in the Amazonian forest, was grown intensively in plantations in Malaya, which dominated world production and fed a growing industrial demand, particularly for vehicle tyres. Beef cattle – not native to the Americas – were reared in the pampas of Argentina, an industry made possible by the development of refrigeration, so that meat could be exported to Europe. Tea, native to China, was planted in places as diverse as India and Kenya. In all these cases, as output expanded, so did the market.

THE PEAK OF IMPERIALISM

In the closing decades of the 19th century, Western powers seized large areas of the world, especially in Africa and in South-East Asia. They prevailed in most places because of their better communications, improved disease control, and sheer military force, armed with industrial firepower.

At the start of the 20th century, decolonization seemed a distant prospect, in spite of the way that the British, French, Spanish and Portuguese empires in the Americas had largely collapsed between 1775 and 1830. This process had not continued elsewhere, although Spain lost its major remaining colonies, Cuba and the Philippines, in 1898, to rebels supported by American military intervention.

By 1900, Britain's empire covered a fifth of the world's land surface and numbered, mostly in India, 400 million people (one quarter of the total world population at the time). France had an empire, mostly in Africa and Indochina, of 15.5 million square kilometres and 52 million people. Other European powers, such as Germany, Belgium and Italy, had recently established colonies in Africa, where Portugal and Spain continued to hold territories annexed centuries before. Europeans ruled almost the entire continent. Territorial expansion overseas had in the past been mostly motivated by trade, but during the 19th century European powers came to argue they had a 'civilizing mission' to rule over 'inferior' races. This mission frequently relied on armies sent by rail and steamships. Often it also depended on local support, and especially on recruiting local troops.

Within Western states a strong sense of imperial mission overlapped with the profit motive, notably in the search for both markets and for raw materials for the expanding industries of Europe. But that motive came

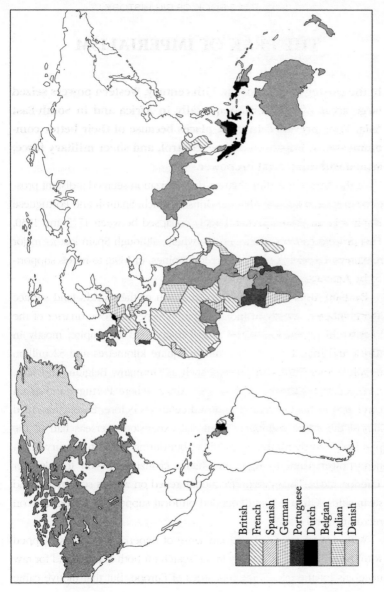

European overseas colonization, 1914

	British
	French
	Spanish
	German
	Portuguese
	Dutch
	Belgian
	Italian
	Danish

second to geopolitics: much imperial expansion came in reaction to the real or perceived intentions of other Western powers, especially in Africa, South-East Asia and Oceania. For example, the British takeover of Burma (Myanmar) in the 1880s was in part designed to stop French expansion.

By 1914, the Western empires had annexed most of the territories they sought and had also defined spheres of influence in most regions still outside the Western grasp, as in China and Persia (Iran). Japan, which had undergone a rapid process of modernization and industrialization from the 1860s, had defied Western control to become an expanding empire in its own right, at the expense not only of China but also of Russia. From both of these it seized large areas on the shores of the Pacific.

But in the 1900s, dislike of Western control and influence led to rebellions in such places as China, the Philippines and South-West Africa (Namibia). These were all crushed, sometimes with great brutality. And yet by 1914 demands for self-government, if not independence, were heard in many colonies, notably India.

Such demands were based on the values of democracy and national self-determination that, at the end of the First World War, were used by the victors to justify the dismantling of internal European empires, such as that of Austria-Hungary. After the Second World War, the same principles, plus the vast economic cost of the war, ended for ever the European empires in Africa and Asia.

'Every empire . . . tells itself and the world that it is unlike all other empires, that its mission is not to plunder and control but to educate and liberate.'

Edward W. Said (*Los Angeles Times*, 20 July 2003)

TRADE UNIONS, SOCIALISM
AND COMMUNISM

In the pre-industrial period, artisans had combined in craft guilds, bodies that restricted entry and set standards within a trade. As industry expanded in the 19th century, the centre of power in society shifted from the landowning aristocracy to the new industrial capitalist class. But a growing class of low-wage factory workers had less independence than the earlier artisans, and worked in often dangerous conditions.

When workers tried to campaign in trade unions for better wages and conditions, governments and employers perceived a threat to property rights and a danger to society. The French Revolution was still a recent memory and there was fear that similar uprisings could happen elsewhere. So unions were often banned, and strikes suppressed violently.

Some thinkers were starting to question the basis of capitalist society. They sought alternatives in which the state would improve living standards and guarantee justice and equality for all. The word 'socialism' was first used by a Frenchman, Henri de Saint-Simon (1760–1825), who believed that industry could remodel humanity without reducing the workforce to poverty.

In 1799 the philanthropist and businessman Robert Owen bought cotton mills at New Lanark in Scotland, which he ran for thirty years as an innovative industrial complex, with good-quality housing and other facilities for the workers. The socialists who succeeded Owen wanted the state to create a better society by providing education, health care, minimum wages, pensions and support in hard times. Some believed that a socialist society could be achieved through democratic reform, others that it would take a revolution.

Communism provided a more radical vision. The idea of a society without class, based on communal ownership of property and wealth, went back at least as far as the early Christian Church. In his book *Utopia* (1516), the English philosopher and statesman Thomas More described a society that held all property in common. Similar ideas were proposed by later groups such as the Levellers (during the English Civil War of the 17th century).

These early displays of communist ideals were often rooted in religious belief, but in the *Communist Manifesto* (1848) the German philosophers Friedrich Engels and Karl Marx took a more materialist approach, arguing that 'the history of all hitherto existing societies is the history of class struggle'. They viewed class as linked to economic power, which derived from the individual's relationship to the means of production. All societies through history had been engines to make goods and to distribute tasks and benefits, and this engine was controlled by the dominant class. In the industrial society of western Europe, Engels and Marx identified two groups: the proletariat or workers lived off the sale of their labour; the industrial bourgeoisie (the capitalists) bought that labour to work in their factories. These groups were in conflict over who should control and benefit from the fruits of labour. The Marxist analysis treated history as a scientifically inevitable process: capitalism had conquered feudalism, and must in turn be violently overthrown by a proletarian revolution. This would lead to the ideal communist society, which would take the means of production into common ownership.

By the end of the 19th century, socialists had begun to make gains on issues such as health and safety regulations and working hours, especially in countries like Britain, where the earlier bans on trade unions had gradually eased. Trade union activism was regarded by both socialists and communists as a crucial part of the struggle against the forces of capitalism.

In the 20th century, communist revolutions in Russia, China and elsewhere created states whose ideology was (at least initially) rooted in the Marxist ideal of the public ownership of industry and land, which would be achieved through nationalization and collectivization. But these regimes fostered the rise of dictators, proved harshly authoritarian, and practised repression as a means of government (see p 214).

By the end of the 20th century communism had been abandoned in Russia and elsewhere, while China had incorporated a far greater level of state capitalism into its economy. In the West, democratic socialist reforms have had a greater impact. In many countries the state retains a strong role in the economy, providing education and health, regulating working conditions and wages, and creating infrastructure.

'From each according to his ability, to each according to his needs.'

Slogan, first used by the socialist Louis Blanc in 1851,
later popularized by Karl Marx

PART SIX

THE MODERN WORLD

The last hundred years have witnessed unprecedented rates of change. There have not only been massive increases in urbanization, industrialization and global population, but staggering advances in science and technology. Humanity has also experienced two of the bloodiest wars in its history, leading to the foundation of new international institutions intended to end conflict. There has also been a growing realization that this small planet is the common home of all humans, and that humans themselves are just one component in the Earth's biosphere.

TIMELINE

1905: Einstein's special theory of relativity.

1911: Chinese revolution ends millennia of imperial rule.

1914: First World War breaks out. Panama canal opens.

1917: Russian Revolution.

1918: First World War ends; collapse of Austrian, German and Turkish empires.

1923: Discovery of galaxies other than Milky Way.

1928: Discovery of penicillin.

1929: Wall Street Crash heralds Great Depression of 1930s.

1933: Nazis come to power in Germany.

1937: War between China and Japan, till 1945.

1939: Second World War begins.

1943: Colossus, first electronic computer, cracks German codes at Bletchley Park.

1945: Atomic bombs destroy Hiroshima and Nagasaki; Second World War ends. United Nations founded. Cold War begins (until 1989).

1947: India becomes independent.

1949: Communists win civil war in China. Formation of NATO.

1950-3: Korean War.

1957: Soviets put first artificial satellite into space. Foundation of Common Market in Europe.

1960: World population 3 billion.

1961: First manned space flight.

1967: First human heart transplant.

1969: First human landing on the Moon.

1975: End of Vietnam War.

1976: US Viking landers touch down on Mars.

1978: First test-tube baby.

1979: Smallpox eradicated.

1989: Invention of the World Wide Web.

1991: Break-up of Soviet Union.

1997: First cloned mammal.

1999: World population reaches 6 billion.

2003: Human Genome Project completed.

2011: World population reaches 7 billion.

2012: Large Hadron Collider demonstrates existence of the Higgs boson.

MODERNISM IN THE ARTS

Modernism is a term applied to a range of international artistic movements that emerged in the early years of the 20th century, with the aim of challenging traditional forms and values.

Modernism drew in part on the new social sciences, notably Sigmund Freud's *The Interpretation of Dreams* (1900). Freud's ideas revolutionized concepts of human behaviour and prompted writers, composers and others to delve deeper into psychological states.

'The inner nature of the unconscious mind is just as unknown to us as the reality of the external world, and is just as imperfectly reported to us through the data of consciousness as is the external world through our organs of sense.'

Sigmund Freud, *Dream Psychology* (1921)

In the second half of the 19th century writers from Baudelaire to Walt Whitman had started to experiment with language and narrative, creating more ambiguous, self-conscious and ironic forms of expression. Modernist literature could be demanding. It challenged the reader with intricate layers of allusion, often to mythology. It also evolved new forms, notably stream of consciousness, which presents the world through the moment-by-moment thoughts and impressions of characters, rather than through the 'objective' voice of the author or narrator. This technique pervades the innovative epic novel *Ulysses* (1922) by James Joyce (1882–1941), where life in modern Dublin echoes episodes in Homer's *Odyssey*.

In poetry, traditional stanza forms, meters and rhyming schemes were replaced by free verse, which, having no predictable structure, continually overthrew the reader's expectations. One of the most influential poems of this period, T. S. Eliot's *The Waste Land* (1922), uses free verse to throw together very different voices and fractured ideas, and, like *Ulysses*, counterpoints myth and literary heritage with the raw realities and alienations of modern life after the devastation of the First World War. Alienation was also explored by Franz Kafka – a double outsider as a German-speaking Jew in Prague, the capital of the new Czechoslovakia. Kafka employed a seemingly realistic narrative style to depict modern life as something akin to a nightmare, in which all human endeavour is doomed to failure at the hands of a faceless bureaucracy administering laws that no one can understand.

Man and machine

'A roaring car that seems to run on grapeshot is more beautiful than the *Winged Victory of Samothrace*.' So wrote Filippo Marinetti, the spokesperson of Futurism, a modernist movement that emerged in Italy before the First World War. Many modernist artists embraced the machine aesthetic, which also became a feature of architecture. Le Corbusier described a house as 'a machine for living in', and modernist architecture discarded the abundant decoration found in many 19th-century buildings in favour of the mantra that 'form follows function'.

Music, too, embraced experiment. Traditional tonality, based on familiar Western scales, was replaced by 'atonality', in which no key was dominant. Arnold Schoenberg in Vienna developed serialism, a new structural convention in which, for example, every phrase must employ all twelve tones and semitones found in the octave. Igor Stravinsky's pounding music for Diaghilev's ballet *The Rite of Spring* caused a riot at its first performance in 1913. The new music that had the greatest public appeal was jazz, born out of the musical traditions of African Americans.

Although its origins were distant from those of classical music, it came to have a great influence on the latter.

Similarly, the visual arts increasingly drew on the arts of peoples from round the world – for example, Pablo Picasso featured African masks in several works painted before the First World War. Western 'high art' no longer possessed a superior status. Picasso also turned new eyes on the physical world, through Cubism. Cubist painters discarded traditional perspective and attempted to depict a person or object from several viewpoints at once, and constructed their subjects out of a series of simpler geometric shapes, such as cubes. The key role of dreams and the unconscious stressed by Freud underlay the rise in the 1920s of Surrealism, in which painters such as Salvador Dalí and René Magritte used meticulously realistic techniques to depict objects and scenes far removed from those encountered in the everyday world. Surrealism also influenced cinema and literature; in the latter, many attempted to create works by surrendering conscious control and employing 'automatic writing'.

For most of the general public, the works of the modernists were for long regarded as incomprehensible, even laughable. In the Soviet Union, after a few years when modernist experiments had been encouraged, Joseph Stalin decreed that modernism in all its forms was an example of Western decadence and bourgeois formalism, alien to working people. Modernism was to be replaced by 'Socialist Realism'. Similarly, the Nazis decried modernist art as 'degenerate' and 'un-German'. Although many elsewhere in the West were similarly suspicious, by the end of the 20th century many aspects of modernism had entered the mainstream, and had come to have a pervasive influence on popular culture.

TOWARDS GENDER EQUALITY

There is some evidence that suggests the existence of matriarchal societies in earlier periods of human history, for instance in the Bronze Age civilization on Crete. In addition the so-called Venus figurines found in Europe from 35,000 to 11,000 years ago clearly grant women high status, and some anthropologists argue that prehistoric hunter-gatherer societies were relatively egalitarian.

Nevertheless, for most of written history, patriarchy (the dominance of males in family and society) has prevailed. But in some parts of the world the position of women began to change from the late 19th century. A crucial development came when women gained the vote: New Zealand came first, in 1893; women in Switzerland had to wait until 1971; women in Saudi Arabia first voted – in municipal council elections – in 2015. Ethnic, class and political biases also operated. When women of European origin gained the vote in Kenya in 1919, for example African women did not. Social criteria included a restriction of enfranchisement in Bolivia to literate women until 1952.

Venus of Brassempouy, carved in mammoth ivory, c. 20,000 BCE

'One is not born a woman: one becomes one.'

Simone de Beauvoir, *The Second Sex* (1949). The idea that gender identity (as opposed to physiology) is a social construct has come to play a vital role in feminist and then in transgender thinking

Although women had toiled through the ages in fields, and later in factories, their work was traditionally different from men's. Late in the 19th century more women began to work in offices, but the professions (such as medicine) were often barred to them. The First World War saw significant changes. Britain was one of a range of countries who mobilized the entire workforce. Women took jobs, for example in munitions factories, that had once belonged to men now conscripted to fight. In both world wars, this phenomenon was much less marked in Germany, where women's roles mostly observed the slogan *Kinder, Küche, Kirche* ('children, cooking, church'). It has been suggested that this failure to fully mobilize the potential workforce was a factor in Germany's two defeats.

Other broader changes in society, including industrialization, urbanization, the decline of deference, secularization and the rise of literacy, greatly affected women as well as men. The extension of state education led to a marked rise in female literacy, and with it a wider range of options, from career choice to social mobility.

One factor among the many that held up women's progress was the partial control they exerted over their bodies. Childbirth grew less hazardous but not safe. Contraception and family-planning advice was limited, often by law. Universities in many countries continued to be predominantly male preserves, as did many professions. Average salaries for women remained lower, and women were denied, or at least did not receive, equal opportunities.

Some of the most important changes came in the later 20th century, with certain scientific advances affecting both sexual choice and safety.

Contraceptive pills and devices gave women greater independence, while antibiotics protected both men and women from the worst effects of some venereal diseases. Some countries granted women equal rights, though action did not always match the laws. In the UK, for example, despite the Equal Pay Act of 1970, average women's pay remains lower than men's.

> **"Let us pick up our books and our pens," I said. "They are our most powerful weapons. One child, one teacher, one book and one pen can change the world.""**
>
> Malala Yousafzai, Pakistani champion of female education and youngest-ever Nobel Prize winner, who survived an assassination attempt in 2012, aged fifteen

Attitudes to sexual behaviour and gender equality have differed between cultures. For example sub-Saharan Africa has seen a growing homophobia, among both Christians and Muslims. Religion has preserved certain sexist attitudes. Unlike some Protestant churches, the Catholic Church has refused to ordain women priests, while fundamentalist Islamic movements are opposed to equality.

Lack of female access to even basic education continues to afflict the developing world. In recent years fundamentalist groups have increasingly used violent means to suppress education for women, and to reimpose traditional gender role in countries from Nigeria to Pakistan.

REVOLUTIONS IN SCIENCE

The early 20th century saw many years of certainty in science overthrown by two new theories: relativity, and quantum theory. Since the time of Newton, scientists had believed that the universe and all it contained could be described in mechanistic terms, obedient to his law of gravitation and three laws of motion. All events – or at least those involving objects and forces – could be counted as predetermined, and therefore predictable.

But although for most practical purposes Newton's laws still ruled, they were now shown to be neither absolute nor universal. According to Einstein's theory of relativity, at speeds approaching that of light, neither time nor mass remain constant. Time and space belong to the same continuum. Both space and light can be bent by gravity.

Again, at the subatomic scale, quantum theory showed that Newton's laws no longer applied. It showed that light, and other forms of electromagnetic radiation, are neither waves nor particles, but *both* waves and particles simultaneously. Other certainties also crumbled. Under Newtonian mechanics the position and momentum of any object can in theory be accurately measured at the same time. Quantum mechanics showed that at the subatomic scale you cannot measure both the position and momentum of a particle simultaneously, because making the observation itself alters the outcome.

These counterintuitive theories – shattering our views of space, time and the very concept of cause and effect – may appear to come from an Alice-in-Wonderland world. Yet many aspects of both have been proved by observation and experiment. Quantum theory accounts for a range of phenomena, from how the eye detects light to the workings of semiconductors, a key technology in modern computers. And Einstein's famous equation, $E = mc^2$ (where c is the speed of light), shows that mass (m) may be converted into energy (E), and is thus fundamental to both nuclear power and nuclear weapons.

The 20th century also saw revolutionary transformations in other fields, from transport, power generation and medicine to agriculture, bioengineering and computing. Scientists provided an explanation of the molecular structure of DNA and hence how genetic characteristics are inherited, and made significant progress in understanding the workings of the brain. As a result of such discoveries, the status of science soared. By the 1950s scientists had become the heralds and guarantors of progress, as engineers had been in the 19th century.

'The unleashed power of the atom has changed everything save our modes of thinking and we thus drift toward unparalleled catastrophe.'

Albert Einstein, in a telegram he sent to prominent Americans (24 May 1946)

A horde of reporters surrounding Professor Albert Einstein in the lounge of the S.S. Belgenland *as he arrived in New York City (1930)*

FIGHTING DISEASE

The 20th century witnessed unprecedented improvements in medical science and practice that touched the lives of billions. Illnesses that had previously been fatal, and diseases that had been debilitating, succumbed to a series of new discoveries.

The discovery of insulin in 1922 enabled many young diabetics to live. Six years later, Alexander Fleming accidentally discovered that a mould called penicillin could destroy bacteria, though it took some years to develop a way to produce it, and to use it against bacterial infections.

Penicillin heralded the antibiotics revolution which began in the 1940s and enabled doctors to treat diseases such as pneumonia, septicaemia and meningitis. At first, antibiotics had no impact on tuberculosis, one of the deadliest killers, as the tubercle bacillus quickly develops resistance to individual drugs. However, in the 1950s it was found that TB could be successfully treated by a combination of antibiotics administered over a long period. Over recent decades, however, there has been an increase in strains of tuberculosis that are resistant even to a combination of antibiotics.

Other bacteria, such as *E. coli*, have also developed resistance to antibiotics. Drug resistance is a growing threat to human health, and could make relatively straightforward surgical procedures much more dangerous, as any consequent infection of the wound may prove not to be treatable. Further threats stem from the routine use of antibiotics in agriculture – for example in China's intensive pig farms – in order to keep animals disease-free and to increase their weight. No new class of antibiotics has been discovered since 1987.

Antibiotics are not the only defences against disease. Common childhood diseases such as measles, whooping cough and diphtheria were effectively controlled by immunization programmes as part of a widespread attempt to improve public health. Vaccination against smallpox was so successful that the World Health Organization declared it eradicated in 1979. A vaccine against polio first went into action in 1956, and this disease too may be eradicated in time. Other viruses, including influenza, HIV and Ebola, have proved harder to combat and continue to challenge medical science, in spite of some progress in the use of antiviral drugs.

Thanks to antibiotics, greater knowledge and improved anaesthetics, exacting operations such as appendectomies became minor routines. New techniques such as blood transfusions, artificial hip and knee implants and the transplantation of human organs became normal. The first human heart transplant was performed in 1967. The manipulation of genes developed from the 1970s and led to the prospect of developing gene therapy for humans.

Mental illness affects a growing number of individuals, but its diagnosis and treatment changed with greater recognition of the role of physiological and neurological processes. The development of safe and effective drugs later in the 20th century helped with major psychoses and depression, as did various methods of psychotherapy, dramatically improving the cure rate. But in much of the world, low levels of institutional provision for mental health necessitate reliance on family networks of support.

As well as vaccination programmes, public health measures have included improved information programmes (even on such simple things as regular hand washing with soap), the distribution of free condoms, and the clearance of stagnant water to reduce mosquito populations. Public health debates encompassed contentious issues of personal, corporate and governmental responsibility. There is for instance much ongoing debate about what, if any, role the state should take in addressing the problems caused by smoking tobacco, by the abuse of alcohol and other drugs, and by the rise in obesity, a trend that is exacerbated by rising affluence and changing social norms.

In the 20th century, the medical profession also took some very different directions. Plastic surgery, developed after the First World War to treat the victims of war, came into common use as an elective, cosmetic procedure. More recently the prescription of performance-improving drugs has become notorious in sports – such as football and athletics – that are also worldwide, billion-dollar industries.

THE ROAD TO WORLD WAR

Few people in 1900 conceived that a major war was coming that would last for years and devastate the old world order. There was tension between the great powers, certainly, and an international arms race was underway, but such tensions were not new, and had not in the past led to major wars.

Europe had been largely at peace since the end of the Napoleonic Wars in 1815. When fighting did break out between leading states – Austria and Prussia in 1866, France and Prussia in 1870–71 – it had all been over in weeks or months.

By the start of the 20th century a new mood of belligerent aggression had arisen among the ruling elites. At this time, conflict between nations was often seen in Darwinian terms, a matter of 'the survival of the fittest'. It was widely believed that the state had a call on its citizens' lives, as embodied in the practice of conscription (first introduced on a mass scale in France a hundred years before). In many European countries young adult males were obliged to serve in the military, generally for two years, followed by an annual service of several weeks in the reserves. This system laid the base for the vast armies that were deployed when the

First World War broke out in 1914. A habit of military obedience colluded with the mood of patriotic war fever to override doubts about going to war, such as religious objections or a belief in international solidarity among workers. Industrialization and the faster pace of technological development could turn out vast arsenals of modern weaponry – including machine guns, aircraft and submarines – to equip these forces. The naval and military spending of the main European powers had doubled in the last two decades of the 19th century, and doubled again in the first decade of the 20th.

By 1914, the leading European powers had formed a series of military alliances. Created as deterrents, to keep the peace and preserve the balance of power, they actually increased the risk of war, because a threat to one posed a threat to all, and the more aggressive members could call the tune. In 1914, as a crisis swelled in the Balkans, Germany and France failed to curb their chief allies, Austria and Russia.

An additional danger was that military planning was entrusted, not to civilian leaders, but to each country's general staff – men who took it for granted that striking the first blow was likely to be decisive. In the run-up to the First World War, their planning revolved around seizing the initiative in order to dictate the dynamic of events.

'We want eight, and we won't wait.'

This slogan emerged in Britain in 1909, in the lead-up to the First World War, at a time when the major powers were competing to build faster and more heavily armed battleships, such as the British *Dreadnought* launched in 1906

INDUSTRIALIZED SLAUGHTER

The First World War of 1914–18 was the bloodiest war there had ever been, partly because it involved the top three economic powers in the world (Germany, the UK and USA) as well as the leading imperial systems, with all the manpower they controlled. The destructiveness of the weaponry, manufactured on an unprecedented scale, was also crucial.

Another key factor was the inability of the combatants to negotiate a peaceful end to the conflict, an inability repeated in the Second World War. As a consequence, the appalling level of destruction, rather than leading to efforts to negotiate, instead contributed to a determination to devote even more effort to the fight.

The war pitted a coalition of Germany and Austria (the Central Powers) against France, Belgium, Britain, Serbia, Russia (the Allies) and Japan. In turn, each side enlisted new partners, though the Central Powers only gained Turkey and Bulgaria, while the Allies' recruits included Italy, Portugal, Romania and, decisively, the United States (in April 1917).

The immediate cause of the conflict was Austrian aggression towards Serbia. On 28 June 1914 a Bosnian nationalist backed by the Serbian Black Hand secret society assassinated Archduke Franz Ferdinand, heir to the Austrian throne, in Sarajevo, the capital of Austrian-ruled Bosnia. Austria turned down Serbian offers of reparations and instead to go to war, intending to crush the growing nationalism within its empire.

Russia now moved to mobilize its armies, and rejected a German ultimatum to recall them. Germany declared war on Russia and then on Russia's ally France. The decision owed much to the German high command, which feared encirclement by the alliance of France and Russia. The German general staff had prepared a plan to knock France quickly out of the war by bypassing the French fortresses along the German border and invading France through vulnerable Belgium. This action pitched Britain into the war, as a guarantor by treaty of Belgian neutrality.

The opening year saw no decisive victories, but the Germans ended 1914 with major territorial gains in both Belgium and France. So the western Allies were forced to take the offensive, both to recover ground and also to reduce the strains on Russia, which lost heavily at the hands of Germany in 1914 and 1915.

In September 1914 the Germans dug themselves into defensive positions, the Allies did the same and the Western Front was born. Its multiple lines of trenches reached from the Franco-Swiss border north-westward to the North Sea. French and British generals dreamed of making a decisive breakthrough through frontal attacks, but the military technology of the day, notably machine guns and artillery, emplaced in strong systems of trenches and dugouts, strongly favoured the defending side. The massive casualties that resulted were not matched by significant territorial gains. By the end of the war, however, new tactics, such as the use of precise artillery fire integrated with infantry assaults, proved more effective. It was these that helped the Allies to victory in 1918.

'We were very surprised to see [the English soldiers] walking, we had never seen that before. The officers were in front. I noticed one of them walking calmly, carrying a walking stick. When we started firing we just had to load and reload. They went down in their hundreds. You didn't have to aim, we just fired into them.'

German machine-gunner recalls the first day of the Battle of the Somme (1 July 1916)

In the meantime, the war had shed blood on an unheard-of scale. Some of the battles, notably Verdun and the Somme in 1916, were ruinously costly (over 2 million casualties between them). Although they knocked Russia out of the war in 1917–18, the Germans were outfought and outresourced by the Western allies.

The war had led to the development of massive military–industrial complexes, especially for the manufacture of guns and high-explosive shells. Control of the seas was also crucial. Germany waged a long and successful submarine campaign against Allied merchant shipping. At the same time, Allied navies, particularly the British, imposed a tight blockade on Germany, and starved it of resources. Both sides used poison gas against each other, though this was never militarily decisive.

This war appeared far removed from previous ideas of battle as a matter of individual and collective heroism. Instead, it embodied the notion that humans came second to the deadly machinery of war. The war also saw the first significant aerial bombardment of civilian targets, which led to the fear that future wars would see cities destroyed by bombers. This all contributed to a degree of subsequent anti-war feeling. However, in the immediate aftermath and on into the 1920s, many people appear to have accepted that the war had been a necessary burden, a matter of unavoidable duty and sacrifice.

'The statesmen were overwhelmed by the magnitude of events. The generals were overwhelmed also. Mass, they believed, was the secret of victory. The mass they invoked was beyond their control. All fumbled more or less helplessly.'

A. J. P. Taylor, *The First World War* (1963)

VERSAILLES AND ITS OUTCOMES

The First World War ended with a series of peace treaties, most notably the one signed with Germany in 1919, the Treaty of Versailles. In imposing these treaties, the victorious Allies aimed both to punish the defeated powers and to ensure a stable postwar world.

The most obvious failure of the Treaty of Versailles was that it did not prevent the outbreak of the next war two decades later. Even in 1919 the economist John Maynard Keynes, who attended the peace conference, predicted that the harshness of the treaty would lead to financial collapse in Germany and further chaos. Equally sceptical was Marshal Ferdinand Foch, the French supreme commander of Allied forces in the war, who complained in May 1919: 'This is not peace. It is an armistice for twenty years.'

The pace of events put pressure on the peacemakers. The war had ended in 1918 with the fall of the Austrian and German governments and the defeat of the Turks. Nationalists in Eastern Europe and in the collapsed Turkish empire now demanded new states. The peacemakers responded by separating Hungary from Austria (its once great empire shrunk to around its present size) and by acknowledging Poland and Czechoslovakia as independent nation-states. Romania expanded. Serbia became the basis of the new state of Yugoslavia.

Germany did not lose territory on the Austrian scale, but it did lose land to Poland, and was forced to return the conquests made from France in 1871. It also had to demobilize and disarm, and to pay reparations: compensation for the damage its armed forces caused. This damaged the German economy and antagonized its people, to the advantage of extremists like Adolf Hitler. The Nazis exploited it as a way to discredit the Weimar Republic, the democratic government formed in Germany in 1919 to replace the imperial system.

The destruction of the Austro-Hungarian empire left simmering grievances. In Hungary anger persists to this day about areas with Hungarian majorities that were transferred to Czechoslovakia, Romania and Yugoslavia (now Slovakia, Romania and Croatia). Turkey lost many possessions in 1920, in particular the Arab provinces conferred on France (Syria and Lebanon) or Britain (Iraq, Palestine and Transjordan). When the Allies carved the Middle East into artificial nation-states, bounded by arbitrary straight lines, they created at least three long-term problems. They thwarted Arab nationalist ambitions, glossed over sectarian divisions (especially between Shia and Sunni Muslims), and ignored the rights of local non-Arab peoples such as the Kurds.

Prompted by the US president, Woodrow Wilson, the Versailles settlement set up a new international body to oversee the global system and to keep the peace, a 'League of Nations'. But America was a land largely settled by Europeans fleeing persecution or war in their own countries. Long averse to 'foreign entanglements', it reverted to its usual isolationism: the US Senate refused to support the League and America did not join it. Add to this the exclusion of communist Russia and the defeated Germany, and the League's global reach was shackled from the start (see p. 236).

REVOLUTIONS

The violent overthrow of existing political systems was a major feature of global history in the first half of the 20th century. Many countries were affected, most notably China and Russia.

Most revolutions reflected the view that monarchical systems were obsolete, and were blocking the changes required to equip a modern state to deal with the modern world: a brutally competitive international system, demands for domestic reform, and the threat of social disorder. In Japan in the 1860s it had proved possible to reconcile imperial legitimacy with radical modernization, but elsewhere, in the early 20th century, the crises were more acute. It was no accident that in a number of states – among them China, Portugal and Turkey – military figures should head the call for change, as it was failures in defence that eroded and cracked old systems. In China the Manchu dynasty lost face after foreign powers intervened to put down the Boxer Rising in 1900. This led to the fall of the Chinese emperor in 1911–12, after more than 2,000 years of imperial rule. A new republic followed, but subsequent decades saw a collapse of national unity as provinces broke away to form warlord states.

The First World War brought down the Austrian, German and Turkish imperial dynasties. It also sparked the Russian Revolution. Repeated defeats at the hands of Germany redoubled the acute social, economic and political strains that the war imposed on Russia, and they discredited the government and especially the Tsar, Nicholas II, who had taken personal responsibility for the war effort. In early 1917 a moderate republican government overthrew him but pledged to continue the war. Its failure was in its turn exploited by Vladimir Lenin's small but determined group of Bolsheviks, the core of the future Soviet Communist Party. They seized power in the October Revolution, and imposed a totalitarian system.

Lenin destroyed his opponents on the Left, and fought a civil war that defeated the right-wing Whites, in spite of international support that included British, French, American, Canadian and Japanese forces. The Bolsheviks held the key manufacturing centres, and their base in Moscow gave control of the rail system so they could dispatch their resources to where they were most needed.

'The substitution of the proletarian for the bourgeois state is impossible without a violent revolution.'

Lenin (1917)

Still weary after the 'Great War', and lacking coordinated aims, the intervening forces withdrew. Added to that, the White idea of 'one Russia, great and undivided' alienated the various nationalist movements that wanted independence from the Russian empire. In the end, those movements failed in Ukraine, the Caucasus and Central Asia, but succeeded in Finland, Estonia, Latvia, Lithuania and Poland. Minus these latter territories, the new Union of Soviet Socialist Republics (the USSR or Soviet Union) established by Lenin in 1922 was effectively a re-creation of the old tsarist empire.

The Bolsheviks' radical reforms included the imposition of state control of agriculture and industry. The liquidation of the kulak class of prosperous peasants displaced over 5 million households into labour camps or distant areas. Lenin had no interest in democracy, which he saw as a bourgeois affectation. The Bolsheviks were committed to change, especially fast industrialization, which they saw as a means of reinforcing their power.

Two key features of the Soviet Union were indoctrination and the secret police. Those viewed as dissidents or counter-revolutionaries were routinely arrested and either executed or sent to the gulag, a vast network of camps. They worked as slave labour in their millions, and many died. This level of paranoia also led to Stalin's show trials and purges, in which party members, leading Bolsheviks and even top officials of the secret police were forced to confess to crimes against the state and were punished or killed as a consequence.

WORLD ECONOMIC COLLAPSE

In 1637 prices for tulip bulbs in Holland rose crazily high – to the point where houses were swapped for a single bulb – then suddenly crashed. This 'Tulipomania' was arguably the first speculative bubble of the modern era.

As banking and capital investment expanded, there were many such bubbles and slumps, including those following the collapse of the Mississippi Company and South Sea Company in 1720. As the money used to buy into such speculative ventures was often borrowed, when the bubble burst borrowers were bankrupted, lenders lost their money, banks failed, and whole economies suffered downturns.

In the 19th century, the Panic of 1873, caused partly by disastrous investments in railways, triggered an economic slump across Europe and America that lasted at least until 1879, and in some views until 1896. This slump, now known as the Long Depression, was called the Great Depression until the disastrous crisis of the 1930s claimed that title.

The Wall Street Crash and the subsequent slump of 1929 led directly into the global Great Depression of the 1930s. This was a period of declining production and trade, of business failures, a collapse of public finances, and a steep rise in unemployment. These were aspects of a fundamental crisis in the world economic system that also fuelled the move from democracy to totalitarianism that was widely seen in the 1930s.

One of the causes was the massive damage that the First World War inflicted on the pre-war economic system. It had left the USA as a major creditor nation and saddled the chief powers involved with heavy debts. In 1929, relatively small-scale problems in the loan system snowballed into a crisis of confidence in asset values. The decade had seen high levels of speculation, for instance in shares, whose prices had soared. When confidence suddenly collapsed, prices fell rapidly. Loans were recalled and there was a run on the weaker banks, which in turn sapped confidence even in the stronger banks, and thus the crises worsened. As credit available for investment and trade fell, economic activity declined. Primary producers (those producing food and raw materials, such as Australia and Brazil) saw their markets shrink in the developed world. Now these producers were less able to purchase manufactured goods from the industrial countries, and their exports suffered in turn.

'Debates go on to this day about what caused the Great Depression. Economics is not very good at explaining swings in economic activity.'

Eugene Fama, *The New Yorker* interview (13 January 2010)

New knock-on effects kept appearing. All over the world, there was a downward pressure on wages and a rise in unemployment, which rose to nearly 24 per cent in the USA in 1932.

Economic strains hit the weakest industrial sections hardest. In Britain, for example, mining and heavy manufacturing were badly hit. However, some industries did continue to grow. Along with cars, more radios and washing machines were produced, and there was a major expansion in the cinema industry. There was also significant industrial growth in the Soviet Union and Hitler's Germany, in part because the governments focused resources on developing heavy industry and a strong military-industrial complex. However this growth came at the cost of a poor allocation of economic resources, and their over-investment in weapons manufacture had damaging long-term consequences.

Crashes pose problems for politicians, especially those with free-market instincts. Some governments at first tried cutting spending and raising interest rates to protect their currencies. Protectionist measures and tariffs were also reintroduced, as they had been during the Long Depression. In contrast the New Deal, introduced in the USA from 1933 onwards, was partly about using government spending to stimulate growth, as advocated by economists such as John Maynard Keynes. Measures were also put in place to regulate banks more effectively.

Industrial recovery remained fairly weak in most countries, including the United States, until the Second World War boosted both manufacturing and employment.

In 2008 a similar combination of speculation and weakened banks led to a new global financial crisis. Responses were as varied as they had been in the Great Depression. Some countries imposed austerity policies, often at the expense of welfare and local-government spending; others advocated measures closer to the New Deal, or other ways of expanding the money supply. How well the world handles future financial crises may depend on how such problems are dealt with in coming decades.

TOTALITARIANISM

The 1930s saw the rise of a number of totalitarian regimes in different parts of the world, as anti-democratic political movements gained power and ruled in a ruthlessly authoritarian fashion.

A number of states already had such regimes, including Russia under the Bolshevik regime, and Italy after the Fascist leader Benito Mussolini seized power in 1922. In the 1930s, more countries saw violent responses to the strains of the global depression. Support for extremist ideologies also played a role, especially in Germany, where in 1933 the National Socialists (Nazis) under Adolf Hitler came to power. He offered the order that many craved, and his overt hatred and scapegoating of social and ethnic groups helped him to build a large following. Hitler played on widespread German anger with the punitive Versailles settlement, and on the myth of the 'stab in the back' – the claim that their own politicians had betrayed the German military. The Nazis obscured the fact that the politicians who sued for peace in 1918 were obliged to do so because of economic collapse at home and military defeats on the Western Front.

> 'The broad mass of a nation will more easily fall victim to a big lie than to a small one.'
>
> Adolf Hitler, *Mein Kampf* (1925)

In Japan, the democratic promise of the 1920s was replaced by authoritarian militarism in the 1930s. Authoritarian regimes also prevailed in Latin America and in Eastern Europe, where only Czechoslovakia remained a democracy. In Spain, the left-wing democratic government of the republic was overthrown in the civil war of 1936–9 by a right-wing group of generals, led by Francesco Franco. The killing of civilians – about

150,000 in Spain during the war and 50,000 thereafter – was a deliberate tactic that both sides employed. Both Italy and Germany subsidized Franco with armaments and troops, while to a lesser extent the Soviet Union helped the Republicans. The failure of Europe's democracies to intervene sent an unintended signal to Hitler's expansionist ambitions.

Authoritarian government was not new. Totalitarianism was. It would become a defining feature of the 20th century as fascist and communist governments alike set out to rule not just the bodies but also the minds of their citizens. They used the mass media (notably radio and cinema) to peddle strident propaganda, whose messages were backed up by secret police and networks of informers. Brutal punishments, imprisonments and murder were used to keep the population in a state of fear. To the far right in particular, with its notion of inferior races, the national destiny was a military one. As a result, Hitler, Mussolini and the Japanese nationalists were ready to risk, or even initiate, war, although none of them anticipated the extent of the world war that was to come.

In the aftermath of the Wall Street Crash, the USA imposed heavy tariffs on Japan, which already felt isolated in a global system rigged, in its view, by the West. To the Japanese leadership, territorial expansion was the only way to protect their economy and to acquire much-needed natural resources. In 1931 Japan invaded Manchuria, in north-eastern China (see p. 236). In 1937 it launched a full-scale war of conquest against the rest of China, which caused massive damage and millions of casualties, many of them civilians. Although the Japanese captured many key cities, by late 1938 they found that they could not subdue the Chinese and could not continue to advance. Their sense of frustration affected their response to the international situation in 1939–41, and convinced them that supply routes to China must be cut.

In Europe, Hitler had for some years been demanding territorial concessions, on the grounds of protecting German-speaking populations in non-German countries. This culminated on 1 September 1939 with his invasion of Poland. Britain and France had been bowing to Hitler's demands for the sake of peace – the policy of 'appeasement'. Now, in support of Poland, they declared war on Germany. This quickly turned to global war, as both Britain and France mobilized their vast overseas empires.

TOTAL WAR

To this day, the Second World War is the most extreme example of 'total war', in which civilian resources and infrastructure are entirely mobilised as part of the war effort, and are also treated by the enemy as legitimate military targets. For perhaps the first time in history, civilian deaths far outnumbered military deaths, and altogether as many as 60 million people may have lost their lives, with many millions more lost to famine and disease.

Hitler craved a Greater Germany, uniting the many areas across Central Europe that were home to ethnic German populations. He aimed to gain *Lebensraum* (living space) for Germans in western Russia, and to conquer further areas to the south for economic resources such as oil. At first he was hugely successful, seizing a large part of Europe between 1939 and 1941. However, despite driving British forces from the European mainland, and also conquering much of the European parts of the Soviet Union in 1941–2, he was unable to end the war.

The new British government under Winston Churchill was not interested in a compromise peace, so the conflict went on until the actions of the Soviet Union and the United States could play a decisive role. Crucially, the Soviet regime did not respond to serious battlefield defeats following the German invasion of 1941 by offering peace terms, as it had done in 1918, partly because Hitler had made it clear that his aim was to either enslave or annihilate all the non-'Aryan' populations to the east, most of whom were Slavs, regarded by the Nazis as *Untermenschen* ('sub-humans').

In the East, Germany's ally Japan won major victories at the expense of the British, French, Dutch and American empires in South-East Asia and the Pacific in 1941–2. Like the Germans, it was never able to call a halt and capitalize on its gains. The surprise attack on the American Pacific Fleet at Pearl Harbor on 7 December 1941 that brought the USA into the war ruled out any remaining chance of a limited war. The USA and its Allies demanded nothing less than the unconditional surrender of Germany and Japan.

The superior industrial and manpower resources of the Allies proved to be decisive. The USA was by far the world's biggest economy, while in the Soviet Union industrial production, particularly of arms, was relocated safely to the east, beyond the Ural Mountains. In addition, Stalin could muster millions of infantrymen to launch wave after wave of counterattacks. By 1945 the Soviets had driven westward into Germany itself. On D-Day, 6 June 1944, Anglo-American-Canadian forces mounted the biggest sea-borne invasion of all time, to land on the Normandy beaches of France and push east towards Germany. Meanwhile, Allied bombers targeted German cities, regardless of their military importance, leaving hundreds of thousands of civilians dead.

On the other side of the world, the Americans drove the Japanese back across the Pacific while the British blocked the Japanese attempt to invade India. In 1945, as Soviet forces fought to enter Berlin, Hitler committed suicide. Germany surrendered in May. The Japanese staged a fierce resistance in the Pacific, and might have fought harder still had the Allies invaded their home islands. However, the situation changed utterly after 6 and 9 August, when the Americans dropped two atomic bombs on the cities of Hiroshima and Nagasaki, with death tolls in single explosions of around 70,000 and 35,000 lives respectively. Japan surrendered six days later.

The US development of the atomic bomb cost billions of dollars and drew on immense scientific, technological, industrial and organizational resources. It was an example of how total war meant mobilizing whole societies, not just men of fighting age, as governments increased their power and directed economies as never before. Conscription was extended, while women were recruited into workforces to an unprecedented extent, with effects felt long after the end of the war. Governments also had to maintain social cohesion and morale, employing methods such as propaganda and police surveillance.

> 'I ask you: Do you want total war? If necessary, do you want a war more total and radical than anything that we can even yet imagine? . . . Now, people, rise up, and let the storm break loose.'
>
> Josef Goebbels, Nazi propaganda minister, in February 1943,
> after the tide had turned against Germany

In many ways, the Second World War transformed the world. The scale of the conflict had been far greater than in the First World War, as had the damage to civilian life. Millions of refugees were left trying to find shelter and new homes in the aftermath. In terms of geopolitics, the United States and the Soviet Union became the world's two post-war superpowers, while the European imperial powers were greatly weakened, losing most of their overseas possessions over the next two or three decades.

Perhaps the most enduring legacy is that the idea of 'total war' has been normalized. The wars and civil wars fought across much of the world since 1945 have seen many wholesale attacks on civilians committed as instruments of policy, the widespread use of rape as a means of punishment or repression, and the enlistment and indoctrination of children to fight.

GENOCIDE

The Holocaust of the Second World War, in which the Nazis meticulously organized the extermination in specially built 'death camps' of two-thirds of the Jewish population of Europe, was the largest-scale genocide in history.

In addition to the murder of some 6 million Jews, the Nazis also killed nearly 400,000 Gypsies, plus large (but unknown) numbers of Slavs, disabled people, homosexuals and political opponents. Some 3 million Soviet prisoners of war also died of starvation, disease and neglect.

There had been genocides before, and there have been genocides since. But genocide – the large-scale extermination of populations on racial, ethnic, political, cultural or religious grounds – is particularly associated with the 20th century. In the ancient and medieval worlds, it was certainly not uncommon for the entire population of a besieged city to be put to the sword if the city failed to surrender. But the attempt to systematically eradicate an entire group of people, usually on ideological or racial grounds, is a relatively recent phenomenon. Examples include the Armenian massacres carried out by the Turks during the First World War; the mass killing in 1971 of Bengalis by West Pakistani forces in what was then East Pakistan as the latter strove for independence as Bangladesh; the 'ethnic cleansing' of Bosnian Muslims by Bosnian Serbs in the 1990s; and the Rwandan genocide of 1994, in which perhaps as many as a million of the minority Tutsi people were slaughtered by the majority Hutu.

In preparation for the Rwandan genocide, extremist Hutus began to refer to the Tutsis as 'cockroaches'. This depiction of intended victims as vermin is a common phenomenon in genocides. For example, the Nazis frequently depicted Jews as rats, and all their victims as *Untermenschen* ('sub-humans'). It seems that people can only be persuaded to participate in, or at least condone, large-scale killings once they begin to think of the victims as completely alien and 'other', indeed that they are not even human.

> 'Anyone who has the power to make you believe absurdities has the power to make you commit atrocities.'
>
> - Voltaire, *Questions sur les miracles* (1765)

THE NUCLEAR AGE

The atomic bombs dropped on Hiroshima and Nagasaki appeared to usher in a new age in which whole cities could be destroyed in a moment, and humankind might come to destroy itself. As it turned out, although the United States enjoyed a monopoly of nuclear weapons from 1945 until 1949, it didn't use them again.

During the Second World War, fearing that the Nazis might be developing such weapons, the USA embarked on its own programme. The resulting Manhattan Project deployed an international team of experts, and led to the production of the A-bombs used against Japan.

In the early years of the Cold War (see p. 224), the nuclear strength of the USA allowed it to demobilize rapidly after 1945, and to act freely around the world. This period of confidence ended in 1949 when it became clear that the Soviet Union had also been able to develop its own atomic weapons, an achievement that reflected not only Soviet advances in science and technology, but also the effectiveness of its espionage networks in the West.

The Soviet success disconcerted the Americans, and encouraged them to develop a much more powerful nuclear weapon, the hydrogen bomb. They tested the first device in 1952, on Eniwetok atoll in the Pacific, and pictures of the fireball, 5 kilometres wide and rising 17 kilometres high, astonished and terrified the world. The following year the Soviets tested their own H-bomb.

Other powers, first Britain then France and China, developed their own A-bombs, and then H-bombs. However, the key powers were still the United States and the Soviet Union, who built up massive arsenals. At first, nuclear weapons took the form of bombs designed to be dropped from aircraft, but by the late 1950s both sides had developed long-range missiles to deliver nuclear devices, either ground-launched or fired from submarines. These missiles moved much faster than aircraft and would be far harder to track or intercept.

The two powers amassed formidable arrays of long-range weapons. This escalation was inspired by the idea that only a massive arsenal could deter the other side from attack. The theory of mutually assured destruction ('M.A.D.') suggested that where two sides both had overwhelming nuclear arsenals, a first strike by one would obliterate both. Planners identified targets that would have resulted in the deaths of hundreds of millions. Civilians grew used to emergency drills and procedures that would have defended them from neither blast nor radiation. Children were even taught to take shelter under their desks at school.

Whether or not these nuclear arsenals prevented full-scale war is an open question. In 1955, President Dwight D. Eisenhower of the United States warned his Soviet counterpart of the risk that such a war might wipe out human life in the northern hemisphere. Eisenhower abandoned the policy of the 'rollback' of Soviet power in Eastern Europe precisely because he feared nuclear devastation. Instead, the superpowers fought smaller-scale but hugely costly indirect wars with each other's allies, as in Korea and Vietnam, but withheld the full range of their forces. Eisenhower did, however, threaten the use of nuclear weapons in order to bring the Korean War (1950–3) to an end after Chinese troops had intervened. Their use also briefly looked imminent during the US–Soviet standoff in the Cuban Missile Crisis of 1962, after the Soviets deployed missiles on Cuba, close to the continental USA. In the end, the Soviets withdrew their missiles.

'I am become death, the destroyer of worlds.'

J. Robert Oppenheimer, the scientist in charge of developing the first atom bombs, on witnessing the first test explosion in the desert of New Mexico in 1945.
He was quoting from the ancient Hindu text the *Bhagavad Gita*.

In the 1970s, interest in nuclear-limitation agreements developed as part of détente (the cautious improvement of relations between the USA and the Soviet Union). After early agreements, relations cooled between the two superpowers, but they moved on to new and more comprehensive agreements from the late 1980s, and these helped to bring about the end of the Cold War (see p. 226).

Instead, the focus shifted to fears about nuclear proliferation – the danger that other states, including aggressive 'rogue' states, or even terrorist groups, could gain a nuclear capability. In addition to the USA and Russia (as it is now), the list of nations with nuclear-weapons capabilities now includes China, France, Britain, India, Pakistan, North Korea and, allegedly, Israel. The future of proliferation is unforeseeable, and there is also the worrying prospect of the use of bacteriological and chemical weapons, or of a low-tech 'dirty bomb' combining conventional explosives and radioactive material.

The nuclear age also saw the development of nuclear energy as a peaceful tool, used to generate power and seen at first as a modern, less

In 1952 the USA carried out the first test explosion of a hydrogen bomb on a remote Pacific atoll.

polluting alternative to coal. However, a series of accidents, particularly an explosion in 1986 in the reactor at Chernobyl in Ukraine (then part of the USSR), led to concerns about potential dangers from leakages of radiation. Such fears were revived by the earthquake that damaged the Fukushima nuclear power station in Japan in 2011. The dilemma remains that nuclear energy offers a relatively benign environmental prospect if it can be made safe. Quite how problematic that 'if' is continues to be the subject of controversy.

THE COLD WAR

From 1945 to 1989 international power politics were defined by the stand-off between a communist bloc led by the Soviet Union and an anti-communist bloc led by the United States. This confrontation was military, political, ideological, cultural and economic.

The Cold War spanned the world and, with the race to land the first man on the Moon, even extended into space. It was motivated by incompatible ideologies and views about the the best way for humanity to flourish. Communist commentators presented an image of Soviet-led equality as the standard of progress, while opposing voices argued that communism was inherently totalitarian, and capitalism was the true path to freedom. Both sides were paranoid: Americans feared a domino effect that would turn more and more states communist, while Stalin's control of Eastern Europe, which was partly a revival of the Tsarist ideal of a Greater Russia, also reflected a desire to create a military buffer zone to protect the 'motherland' from any more invasions like that of the Nazis, which had killed some 20 million Soviet citizens.

The end of the Second World War, in which the West and the Soviet Union had cooperated to defeat Nazi Germany, had broadly divided Europe into areas that had been liberated by Soviet and by Western Allied armies respectively. This division became known as the Iron Curtain, and took the form of a heavily militarized line that divided Western and Eastern Europe for over four decades. In 1945, Germany, and its capital Berlin, were divided into military occupation zones with the Western Allies (Britain, France and the USA) controlling the west and the Soviets controlling the east. West and East Germany became separate states in 1949, while the division of Berlin was reinforced in 1961 with the building of the Berlin Wall.

At the Yalta conference in February 1945, the Western Allies had effectively granted Eastern Europe (bar Greece) to the Soviet Union as a sphere of influence. Elsewhere, communists tried but failed to take power in Iran, Greece, Malaya and the Philippines. Soviet expansionism led in 1949 to the foundation of the North Atlantic Treaty Organization (NATO), a defence alliance of a number of North American and West European countries intended to combat further Soviet advances in Europe. The USSR responded with the Warsaw Pact of 1955, which allied the Soviet Union with Albania, Bulgaria, Czechoslovakia, East Germany, Hungary, Poland and Romania.

In China, the communists under Mao Zedong won the civil war of 1946–9. The Japanese had occupied Korea, and the postwar settlement divided the peninsula. The Korean War of 1950–3 began when the communist North Korea invaded the US-allied South. A large-scale US-led United Nations military commitment drove back the North and its Chinese allies, and ended with an armistice that has never become a peace. In this fraught climate, military expenditure rose greatly in North America and Western Europe in the early 1950s.

'Every gun that is made, every warship launched, every rocket fired signifies, in the final sense, a theft from those who hunger and are not fed, those who are cold and are not clothed. This world in arms is not spending money alone. It is spending the sweat of its labourers, the genius of its scientists, the hopes of its children.'

President Dwight D. Eisenhower (16 April 1953)

The Cold War witnessed many other confrontations and conflicts, from the Vietnam War and the nuclear arms race to wars in the Middle East, sub-Saharan Africa and Central America. These regional wars were all at least partially products of the Cold War, although there were also more local issues at stake. In Vietnam the USA intervened in a long proxy war but failed to prevent victory by the communists. The consequences of failure were lessened by a diplomatic realignment in the early 1970s that led to cooperation between America and China and marked a further weakening in Soviet power, following the ideological break between the USSR and China in the early 1960s.

Tensions flared again between the USA and the Soviet Union following the Soviet military intervention in Afghanistan in 1979 and the suppression of a popular reform movement in Poland in 1981. President Ronald Reagan launched a massive increase in US spending on projects such as the neutron bomb, and his 'Star Wars' satellite defence system.

From 1985, however, tensions eased under a new Soviet leader, Mikhail Gorbachev. Gorbachev saw that the Soviet Union could not possibly match its rival's military spending and that the Soviet economy was becoming overstretched. However, his policies of *glasnost* (openness) and *perestroika* (restructuring), inadvertently led to the fall of communist regimes in Eastern Europe in 1989 and finally to the collapse of the Soviet Union itself in 1991. Once Gorbachev let it be known that the USSR would not intervene militarily in the affairs of Eastern-bloc countries, the communist regimes there were left unable to resist popular pressure for change. Huge crowds demonstrated for change, and in November 1989 the opening of the Berlin Wall enabled East Germans to flood into the West. This sign that the East German regime was tottering brought other popular movements onto the streets of Eastern Europe. The resulting transitions to non-communist governments were largely peaceful, except in Romania, where the regime violently, but unsuccessfully, resisted change. By contrast, in China, the communist government retained control. A mass pro-democracy movement had been suppressed with much bloodshed earlier in 1989. But the Cold War had come to an end.

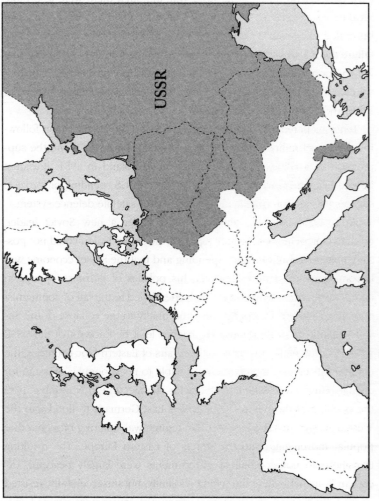

The USSR and its communist allies in Europe, 1950

LIFE AFTER THE COLD WAR

The fall of the Soviet Union in 1991 inspired much talk of a new world order, headed by the USA and reinforced not only militarily, but by the global spread of the American model of economic liberalism with its commitment to free markets, deregulation of finance, and the privatization of state-owned assets.

But the USA ran into severe opposition in the early years of the 21st century. In the Islamic world, America and its allies clashed with a range of opponents. Religious fundamentalists often adopted terrorism as a strategy, most dramatically with the attacks on New York and Washington on 11 September 2001 that used hijacked passenger aircraft as weapons. Fundamentalists exploited popular hostility to what they called Westernization, which they presented as a form of globalized 'Crusader' power hostile to Islam. The initial response of the USA to the 9/11 attacks was to start wars in Afghanistan and Iraq, and it later backed popular uprisings in other parts of the Middle East. Neither tactic brought peace, and the backlash both inside and outside the region has forced the USA to re-evaluate its military and diplomatic response to a world of 'asymmetric instability', in which its main enemies might be small armed groups rather than superpowers.

At the same time, other great powers were becoming more assertive. China's economy, mostly based on manufacturing, was expanding fast. Nominally communist China had adopted those features of capitalism that it chose, though unchecked by democratic control, as it remained an authoritarian one-party state. Meanwhile in Russia, the upheavals of the end of communist rule had begun to subside. These power shifts weakened the USA's ability to confront China's growing territorial ambitions in the East China Sea, or Russia's renewed desire to dominate countries that had formerly been part of the Soviet Union.

The economies of other nations were also evolving. For instance, India's growth rate overtook China's in 2015, with an economic and political approach that was less economically liberal and more corporatist than the American one.

To focus solely on the tension between the United States and its potential rivals risks downplaying other sources of friction worldwide. One common factor, though, was the use of violence to secure political outcomes. Countries all across the Middle East, Europe and Africa experienced civil wars and internal conflict. In the Democratic Republic of Congo alone, civil wars have cost more than 5 million lives since the 1990s. This conflict combined bitter ethnic differences with interventions by neighbouring African powers, as well as intense competition to control valuable natural resources. From these perspectives the world of the 21st century looks depressingly like the world of the 20th century, one in which military conflict and instability are ways of life for many of the world's population.

BRICS

The late 2000s and early 2010s brought the prospect of a new pecking order in the world economy, as a number of countries began to play more prominent roles. The most significant were the 'BRICS' countries – Brazil, Russia, India, China, South Africa – and after these the 'MINT' countries – Mexico, Indonesia, Nigeria, Turkey. The USA and Europe appeared to be in crisis, while these other countries prospered. By the mid-2010s, however, Chinese growth was slowing, while Russia suffered a major loss of revenues caused in part by the plunging price of oil and in part by sanctions incurred after its interventions in successor Soviet states. Such developments cast doubt on the futures of many economies and states, including the oil-backed systems of the Middle East.

THE INFORMATION REVOLUTION

The distribution and sheer volume of information has expanded massively over the last 500 years. Each new technology – from the early days of printing (see p.154) to mass-produced books, and on through telegraphy, the telephone, radio, television, communications satellites and computers – had huge repercussions.

The 21st century kindled a new information revolution as the Internet magnified the impact of computers and delivered once unthinkable volumes of information to anyone who owned a mobile phone or other portable device.

Computing methods have been in use since before the Second World War, but the need to break enemy codes (particularly the German Enigma code) during the war gave a particular boost to computing theory and computational machines. The early computing devices had specialized tasks to perform, and it was not until 1946 that Americans built the first general-purpose computer. It was funded by the US army, and military needs have funded many advances in computer technology since then.

At first, computers were bulky, expensive devices. The biggest changes came when innovations such as the silicon microchip made computers, and computerized devices, small and cheap enough to acquire a mass market.

From the late 1970s, computers became widely available as office and then household tools. The Internet, which had its origins in 1960s research into robust computer communications networks, took a step forward in the 1980s with the development of the Internet protocol suite (the networking model that allows computers to communicate information). The invention of the World Wide Web in 1989 opened up the Internet to many more people, as well as enabling the growth of email, which for many users has entirely replaced sending letters through the post.

Improvements in network computing enabled interlinked machines to work as a single, much more powerful machine, removing the expense of a supercomputer. Developed in the 1990s, this technique anticipated the later 'cloud-computing' method by which large numbers of machines were combined. 'Cloud computing' drew on the processing power that exists in the 'cloud' created by the general use of computers, and does not require their physical presence. This practice gave small computers a boost: miniaturization was crucial in popularizing new consumer goods such as mobile phones, laptop computers and portable media players.

'Cyberspace is where a long-distance phone call takes place. Cyberspace is where the bank keeps your money. Where your medical records are stored. All of this stuff is out there somewhere. There is really no point in thinking about its geographical location. Information is extra-geographical.'

William Gibson, US science-fiction writer, interviewed in 1995. Gibson had first used the term 'cyberspace' in a 1982 short story to denote 'the mass consensual hallucination' of computer networks

The information revolution draws on cultural and commercial factors as well as technological advances. Demand for such products derives from greater literacy and greater wealth, as well as lower costs of production. World literacy levels, low in 1900, have steadily risen since. Greater average per capita wealth has also made it easier to acquire the new devices, and the world of data they contain. This rise in wealth was particularly apparent in China and India, but was also seen in other areas where the use of new technology grew, such as East Africa.

On top of the flood of information that the touch of a button now summons, there have been exponential rises in computing power, speed and data-storage capacity. More than 2 billion people around the world have access to the Internet. Every minute, YouTube users upload around 50 hours of video, while Facebook users share about 700,000 items of content. Future advances may expand 'the Internet of things' as more and more devices are connected to and operate via the Internet, from cars to heart monitors and home appliances. A growing proportion of the economy is devoted to the development and marketing of computer-related goods and services that no one would have dreamed they would ever want or need a decade or so ago.

Controlling the internet

The spread of information technologies drew new attention to a host of issues about open access and control, especially as the Internet became a more political medium. Virtual communities were created without regard to distance, national borders or social mores, and the Internet became a focal point for a range of human interactions, from dating to political agitation.

Those states that want to suppress such new freedoms tend to focus on surveillance and censorship. At the same time, there is genuine concern about the weakness of states in the face of terrorist and similar threats, whose early signs might be detected by monitoring emails and social media postings.

The conflict between protecting the security of the public and the privacy of the individual will clearly continue into the future. Governments are already seeking to access encrypted information, while technology companies seek to offer privacy to those who demand it, and ordinary users face dilemmas about how best to respond to these challenges.

THE PROMISES OF BIOSCIENCE

Humankind has been reshaping other species around the world for thousands of years. We have used selective breeding to create varieties of plants and animals that have the features we desire, whether it be juicier apples or fatter sheep. In the 19th century, the Austrian friar Gregor Mendel (1822–1884) established the basic rules of heredity through his experiments with pea plants, and this led to a clearer understanding of the scientific basis of selective breeding.

Once Mendel's work became more widely known, it transformed the agriculture of the 20th century, leading to what was termed a 'green revolution'. Improved crop strains – combined with mass-produced chemical fertilizers and pesticides, mechanization and greater use of irrigation – pushed up average yields of grain. In Asia, the most populous continent, the price of rice fell by an average 4 per cent a year from 1969 to 2007 as new strains were developed. Even so, malnutrition is still widespread in the developing world, although often the cause is the poor distribution of resources, rather than sheer inability to provide enough food.

There are limits to what agricultural science can achieve. In the 1950s and 1960s, the Soviet Union attempted to convert the steppes for the cultivation of cotton and wheat. However its massive irrigation programmes diverted water from the rivers that fed the Aral Sea, shrinking the volume of the world's fourth-largest lake to just 10 per cent of its fullest reach, and causing havoc to the region's ecosystems. In many places, the spread of monocultures (where vast areas are devoted to a single crop for the sake of efficiency) led to a fall in biodiversity and encouraged certain pests to flourish. Chemical fertilizers and pesticides increasingly affected the crops that were consumed as well as the water supply, the food chain, and the atmosphere.

Genetic modification of crops brings problems too. Many scientists believe that to genetically alter crops to give them, say, greater disease resistance, or to lower their need for pesticides, has the potential to increase food production and end hunger. Although worries about the potential effects on human health of consuming GM crops have largely been allayed, there are fears that genetic modifications could jump between plant species, with unpredictable consequences. GM research and production is largely done by big corporations, and there are concerns that they might seek to profit from their patented products without considering the more general welfare of humanity.

Genetic engineering as applied to human health is also a matter for research and debate. We have a growing ability to map the genetic code of individuals and to use techniques that alter the genetic make-up of an organism. In 2015 Chinese scientists altered the DNA of human embryos so as to modify the gene responsible for the fatal blood disease thalassaemia. The recent science of epigenetics studies heritable changes (sometimes with environmental origins) that affect the way genes work – their 'expression' – but do not change the DNA itself. Researchers keep uncovering the role of epigenetic factors in all sorts of human disorders and fatal diseases. This may lead to progress, for instance, with some types of cancer.

'The genetic manipulation of plant or animal species enables companies, by enforcing industrial patents, to become owners of all the modified plants and animals subsequently produced . . . A firm can become the owner of an entire species. It's the logic of industry, applied to life.'

José Bové, French Green politician, *The World is Not For Sale* (2002)

Cloning produces an animal that is genetically identical to one of its parents. Developments in cloning technology in the early 21st century it brought both new possibilities and ethical dilemmas. By 2015, China had plans for a cloning factory to grow large herds of beef and dairy cattle. At the same time, the European Parliament imposed a ban on cloning animals for food. Human cloning is even more controversial. Reproductive cloning would create a whole human being from cloned cells. It is altogether banned in many countries. Therapeutic cloning, which reproduces individual cells, has potentially lifesaving medical applications, but is opposed by those who object to the use of stem cells from human embryos on ethical or religious grounds.

Nanotechnology (the 'nano-' part means one-billionth) is another field that may transform future technology and medicine. The creation of functioning machines as small as a molecule or even an atom is still a largely theoretical prospect, though it is the subject of intensive research. It is already possible to create transistors that have a diameter of just a few hundred atoms, and there has been progress in creating and manipulating materials that could be used at a nano-scale. As nanotechnology moves on, its range of possible applications includes diagnosing disease and repairing cell damage.

One dramatic future prospect is the use of artificial implants to restore or even enhance the human body. We already use external devices such as hearing aids and internal or integral ones such as artificial pacemakers, prosthetic limbs and cochlear implants. There has been progress in the creation of brain–computer interfaces, devices that can map, augment or repair sensory functions. 'Transhumanist' thinkers argue that we will eventually be able to transform ourselves into beings with such advanced abilities and functions that we will become 'posthuman beings'.

INTERNATIONALISM, GLOBALIZATION AND THE FUTURE OF THE NATION-STATE

The 20th century saw a number of attempts to establish an effective system of international order, to lessen the risks of war and to settle other international issues. In the previous century the Red Cross and the first Geneva Convention had represented some progress towards agreement on conduct in wartime, but the League of Nations after the First World War (see p. 204) was the first major attempt to establish a pan-national organization with a mission to prevent war and to tackle unresolved issues in peacetime.

Founded in 1919, the League achieved early successes – for instance it dealt with the humanitarian crisis caused by fighting in Turkey in 1922–3. It also sought to act on moral issues: it defined slavery in 1926 and made membership conditional on its abolition. But its response to the Japanese invasion of Manchuria in 1931 and the Italian invasion of Abyssinia (Ethiopia) in 1935 proved ineffectual. The League received little global cooperation. Nations could flout its sanctions and requests for arbitration. One serious outcome was the failure of disarmament talks in the 1920s and 1930s. European powers continued to rebuild their arsenals, and this in turn eroded international faith in the League's ability to prevent future conflict.

'One of my earliest memories is walking up a muddy road into the mountains. It was raining. Behind me, my village was burning. When there was school, it was under a tree. Then the United Nations came. They fed me, my family, my community'.

Ban Ki-Moon, UN secretary-general (2011)

The Second World War marked the ultimate failure of the League, and in 1943, the USA, the Soviet Union, Britain and (Nationalist) China laid the foundations of what was to become the United Nations. However, the UN often proved to be a forum for growing Cold War tensions, not the solution. Other new institutions included the World Bank, the International Monetary Fund and the General Agreement on Trade and Tariffs (GATT). They helped to strengthen the global financial system and to increase global trade. Other UN agencies included the UN Refugee Agency (UN-HCR), the World Health Organization (WHO), which helps to coordinate international efforts to combat dangerous diseases, and the UN cultural organization (UNESCO), created to protect the world's cultural heritage.

One solid achievement is that the UN has made it a normal and regular practice for the world's governments to gather to discuss their concerns and grievances; it thus upholds the concept of international law. But the make-up of the UN Security Council gives any of the five permanent members (China, Russia, France, the UK and the USA) a veto over resolutions. This can be seen as undemocratic and has often led to inaction over crises where permanent members have special interests.

The UN, via its International Court of Justice in The Hague, has been successful in prosecuting war criminals such as the Bosnian Serb leader Radovan Karadžić. Other successes include UNESCO's protection of the fragile ecosystems of the Galapagos Islands. The UN has also, on numerous occasions, acted to lessen the consequences of war and famine. Some of its notable failures have come when its peacekeeping forces failed to prevent genocide in Rwanda in 1994, the massacre of Bosnian Muslim men by Bosnian Serb forces in Srebrenica in 1995, and other such atrocities.

Some have seen international government, through the UN, as the future, but the reality is that major states like China, Russia and the USA continue to follow their individual interests and have favoured internationalist policies only when it suited them. This means that many international issues and conflicts are still dealt with through co-operation between individual nation-states.

However, nation-states face significant challenges to their sovereignty. The power of national governments is restrained by factors beyond their own borders. International blocs such as the European Union originated as free-trade areas, but have transformed into organizations whose laws and regulations their members must obey. Most states have signed international agreements that are binding on their own governments.

Economic globalization, which places large companies outside the national context, also erodes the power of the nation-state. They can avoid paying taxes in the countries where they make much of their income. And they can refuse to invest in (for example, by building factories), or can abandon countries in which they believe the policies of democratically elected governments are going to diminish their profits. Culture too has become globalized, owing to rising ease of communication, the scope of mass media (film, TV and popular music, for example), and the way that social media builds transnational communities of common interest. Whether the institution of the nation-state – a relatively recent form of human social organization – will survive in the long term is an open question.

'The central challenge we face today is to ensure that globalization becomes a positive force for all the world's people, instead of leaving billions behind in squalor.'

Kofi Annan, UN secretary-general (April 2000)

Universal human rights?

The concept of human rights can be traced back at least to the Enlightenment (see p. 162). The idea that individual citizens have rights that they only submit to the rule of government for the sake of the general good underlies the idea of the social contract (see p. 168). Such ideas inspired both the American and the French Revolutions, and in both the revolutionaries drew up statements of citizens' rights.

In 1948 the United Nations adopted the Universal Declaration of Human Rights, and it went on to produce further statements of rights. In 1953 the Council of Europe, much larger than the European Union, launched the European Convention of Human Rights, and citizens in all forty-seven member states can appeal to the European Court of Human Rights. In such documents, rights typically protect individuals against government interference, and guarantee freedom of expression, a fair trial, privacy, a family life, and equality. Other rights, such as those to freedom from hunger and to education, may require positive action by governments.

Critics, sometimes known as 'cultural relativists', have attached the idea of universal human rights as an instance of neo-imperialism, in which Western democrats and liberals impose their values on the rest of the world and ignore local traditions and customs.

Opponents of this viewpoint would have every human treated the same. For example, they might point out that denying or preventing equality between the sexes, as many societies do around the world, simply protects a patriarchal power structure, and is no more defensible than slavery.

POPULATION

The massive and seemingly inexorable rise in world population over the last century has tested the complex relationship between resources and demands.

This trend has been most stark in recent decades. The world's population is projected to reach 8.1 billion by 2025 and 9.7 billion by 2050. Factors causing this unprecedented rate of growth include lower death rates at birth and during infancy, and greater life expectancy. These in turn reflect better living conditions and health care. And population growth is exponential: if a couple have more than two children, and each of these goes on to rear more than two children, then populations will grow at an increasing rate.

The spread of effective contraceptive methods since the mid-20th century has enabled couples to plan their families. Typically, in countries

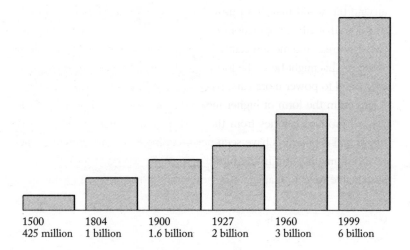

| 1500 | 1804 | 1900 | 1927 | 1960 | 1999 |
| 425 million | 1 billion | 1.6 billion | 2 billion | 3 billion | 6 billion |

with better education and higher standards of living, couples will opt for fewer children, and populations flat-line or even decline. In poorer countries, where people have less chance of education, and may look to their children to support them in their old age, populations typically rise.

There have been a few attempts to impose control: India ran a forced sterilization campaign in 1975–7, and China restricted families to one child until it discarded this policy in 2015. A key issue in the developing world is women's education: where it is available, women tend to marry later and bear children at a later age.

In those countries where birth rates have been falling or static, there are concerns that there will not be a sufficiently large workforce to support an ageing population. This has been a major issue in Japan, for example. In some states close to 'ZPG' (zero population growth), for example Hungary and Italy, population has risen only, or largely, due to immigration.

Early economists used to predict that unbridled population growth would lead to food shortages and inevitable starvation. But many argue that famine is more likely to result from structural economic imbalances around the world than from growing too little food. The fact remains, however, that when populations get richer – as in the West over the past two centuries, and more recently in China – they use more and more resources. This might be in the form of fossil fuels, to generate more electricity and to power more cars, so contributing to global warming. Or it might be in the form of higher meat consumption, which is a wasteful way to produce calories from the land. In recent times most governments and economists have seen never-ending economic growth as the goal of humanity, but in the future we may be forced to look for more sustainable ways to manage our uses of land and of natural resources.

MIGRATION

Rates of migration rose in the 20th century in response to both pull and push factors. The pull factors came in the form of easy air travel and improved land and sea routes, the push factors from armed conflict and political, ethnic or religious persecution, from poverty, and also from natural disasters such as droughts and floods.

Patterns of migration over the last century or so have been varied and complex. Migrant flows *between* countries often drew the most attention, especially where national identities were involved. But much migration was *within* countries, in particular from countryside to city, as the rate of industrialization increased. For many young men, service in the army, whether in war or in peacetime conscription, broke the village connection. Extensive migration within the USA reflected both economic opportunity and patterns of retirement. There were major population shifts from the Rust Belt zones of the North-East and Midwest (where heavy industry was in decline) to the Sun Belt of the West and South-West, as well as to Florida and North Carolina.

Economies varied widely as to how migrants were treated. The huge refugee camps seen in parts of the Middle East and East Africa contrast with the USA, for example, which has attracted large numbers of economic migrants from Latin America. There were similar flows of Turkish workers to West Germany, and Portuguese to France, from the 1950s. By 1973, 12 per cent of the West German labour force was foreign-born. Foreign nationals outnumber the local workforce in the rich but not populous states of the Persian Gulf.

Refugees

The UN's Refugee Convention of 1951 defines a refugee as someone forced to flee their country because of persecution, war or violence, or who suffers a well-founded fear of persecution for reasons of race, religion, nationality, political opinion or membership of a particular social group. Wars throughout the 20th and 21st centuries have led to great flows of refugees. The Second World War alone created an estimated 60 million. In 2004, it was estimated that there were 37.5 million refugees worldwide. By the end of 2014, the figure had risen to 60 million, nearly two-thirds of them internally displaced. Most countries have signed up to the Refugee Convention, and so theoretically recognize their obligations to grant asylum to, and look after, refugees who end up within their borders. At the same time many countries make considerable efforts to prevent refugees from arriving on their shores in the first place.

These migrant workers often do not settle permanently – indeed, the Germans referred to them as *Gastarbeiter* ('guest workers'). This contrasts with those migrants fleeing persecution, who cannot return home. Examples include the Huguenots – the French Protestants expelled from France by the Catholic Louis XIV – who settled in England in the late 17th century, and the Jews who fled Nazi Germany in the 1930s.

As environmental factors and military conflicts continue to stimulate flows of migration, nation-states around the world will face increasing political and economic challenges in absorbing different population groups and cultures. In facing these challenges, we are also reminded that we all share a common humanity, and that we all live on the same small planet, which we happen to have divided up with often arbitrary lines called borders.

ECONOMIC DEVELOPMENTS

Both Western production and consumption shot up after the havoc of the Second World War. The 'Long Boom' was a period of industrial and economic advance and high employment that lasted from 1945 to 1973.

The American economy produced large quantities of affordable consumer durables. It became a society of mass affluence, and Hollywood and television spread positive images of American life. West Germany, Japan and South Korea also saw high economic growth based on strong exports. Technological development had a global impact much faster than in the past, for instance in the development of synthetic fibres such as nylon and polyester. Manufacturing growth changed not only the developed world, but also the developing world, where output rose and diversified.

Economic growth spread wealth and purchasing power. As more and more people left the areas they had grown up in, and found new jobs, new houses, and greater financial independence, they became 'consumers', for whom shopping and fashion could be seen as casual pastimes. A throwaway society discarded goods before they wore out. The Long Boom also fostered ambitious social programmes as rising income levels gave governments higher tax receipts.

More troubled periods followed, with severe recessions in the mid-1970s, the early 1980s, the early 1990s, and from the late 2000s to mid-2010s. In the 1960s, varying rates of inflation around the world had put strains on the international economy, and this led to the collapse of the Bretton Woods system of fixed exchange rates established in 1944. In 1971 the USA ran its first trade deficit of the century. Political crisis in the Middle East led to a dramatic rise in the price of oil in 1973–4 and contributed to 'stagflation', a blend of stagnation and inflation, which intensified uncertainty and malaise.

Yet most peaceful societies still enjoyed a long-term rise in productivity, and the world got wealthier, at least as measured by access to goods and services. Global economic growth was spurred by innovation, especially the creation of new areas of demand, such as personal computers. At the same time, the balance of economies tipped, with a greater stress on services, in sectors from medical care to shopping. The fact that a smaller percentage of the global population could produce necessities such as food and clothing meant that more of the workforce was available for other tasks. Greater average wealth meant that more money was spent on services, and this fed into the growing power of the state to provide its own services.

Within both goods and services, there were also major shifts of emphasis. In industrial production, the share of electronic goods increased, while the more traditional sectors of heavy industry and textiles declined, at least in the West.

There were also major geographical shifts in industry. In 1950, the main industrial areas had been Western Europe, the United States and European Russia, but from the 1960s, industrial output rose markedly in East Asia, first in Japan and then, from the 1980s, in China. China became the world's second-largest industrial producer in the 2000s, and the largest by the 2010s, though more dependent on export than on domestic markets – in 2005, the American trade deficit with China was $202 billion. The problems that hit American manufacturing, as it competed with Chinese and other manufacturers, were reflected in the median income of American households, which more or less stagnated in real terms from 1989 to 2014. In relative terms, high-wage European producers declined except in niche areas such as machine-tools and pharmaceuticals, although, with its accumulated resources, skills and wealth, Europe remained an important economic zone.

'In this new market . . . billions can flow in or out of an economy in seconds. So powerful has this force of money become that some observers now see the hot-money set becoming a sort of shadow world government – one that is irretrievably eroding the concept of the sovereign powers of a nation-state.'

Business Week (20 March 1995)

Two questions stand out among future economic challenges. How will the world square the goal of economic growth with limited natural resources and environmental factors? And how will developing countries balance their economies as expectations of wealth and consumption rise among their populations? Since the international financial crisis that started in 2008, triggered at least in part by irresponsible bank lending and the creation of ever more complex financial instruments, one urgent task for governments is to regulate and police the global financial sector so as to lessen the impact of future financial crises.

ENVIRONMENTAL PROBLEMS

Environmental politics became more prominent from the 1960s onwards. 'Green' movements and political parties were founded in many countries, and many existing political parties adopted pro-environmental policies. More and more people accepted that environmental issues required global action.

Some of the environmental problems that gained attention in the later 20th century had their origins millennia earlier. Deforestation in Europe, for example, had begun with the arrival of agriculture in the

Neolithic period (see p. 80). As Europeans settled in other parts of the world, population pressures led to similar environmental impacts, such as the clearance of large tracts of tropical forest to raise cattle, or for the production of soya or palm oil. In other places forest has been destroyed so that dams can be built to generate hydroelectricity. As well as great reservoirs of biodiversity, tropical forests are huge carbon sinks, which reduce the amount of carbon dioxide (the most plentiful greenhouse gas) in the atmosphere. The build-up of these gases has been rising since the Industrial Revolution, which led to a vast increase in the use of fossil fuels such as coal and oil, whose burning releases carbon dioxide into the atmosphere.

Very often, the damage that human activities do to the environment is not confined to a single country or region. Pollutants are carried around the world by wind and ocean currents. The greenhouse effect is a global one: gases like carbon dioxide build up in the atmosphere and act like the glass panes of a greenhouse stopping heat escaping from the planet. As icecaps at the poles start to melt, there is less ice to reflect the Sun's heat, and the planet absorbs even more, in a feedback cycle. As the permafrost melts in the great northern tundras, more and more methane (another greenhouse gas) is released into the atmosphere. The melting of polar ice also raises sea levels, and threatens to drown heavily populated low-lying areas all around the world. Global warming also leads to destructive changes in climate, from an increase in droughts (and hence desertification) in certain areas to an increase in storms and devastating floods in others.

Concern about global warming led in 1992 to the Earth Summit in Rio de Janeiro, which agreed the Framework Convention on Climate Change. This in turn led in 1997 to the Kyoto Protocol, under which the major industrial countries accepted significant reductions in their emissions of greenhouse gases. But it proved hard to reach agreement on how to enforce the protocol. Among the prime polluters were newly industrializing countries, especially China and India, who felt they should bear a much lighter burden than those countries (for instance in Europe and North America) that had already industrialized. In 2001 the USA, whose emissions rose sharply in the 1990s, rejected the Kyoto treaty. However, in 2015, an international conference in Paris agreed to take further steps to limit global warming.

Other forms of air pollution have also been problematic. Acid rain damages trees, rivers and lakes. It is caused by emissions of sulphur dioxide and nitrogen dioxide from coal-fired power stations and other industrial processes. These chemicals combine with moisture in the atmosphere to form dilute acids (sulphuric and nitric) that fall as rain. In the late 20th century, the lakes and forests of Scandinavia were being badly damaged by acid rain caused largely by industrial activity further to the south, in Germany and elsewhere. Recognition of the problem led to action, for example the installation of filters to reduce dangerous emissions from power stations.

Lead-free petrol, and catalytic converters for car exhausts, also lowered levels of acid rain. But as cars and trucks proliferate worldwide, so do the threats to human hearts and lungs that their emissions cause. Some cities, such as Paris and Delhi, have sought to reduce this kind of pollution by banning drivers from driving on certain days. Other atmospheric threats to human health come from tiny particles produced, for example, by coal-fired power stations, diesel exhaust, construction dust, and burning crop waste. In Europe alone, by 2015 over 400,000 people were estimated to die prematurely each year owing to urban air pollution.

The consumer society also produced greater and greater quantities of rubbish, much of it non-biodegradable and some of it toxic. The proliferation of tiny particles of plastic (often from bottles) in the oceans proved devastating to many forms of marine life. Oil spills from drilling rigs or

A landfill site in Jakarta

damaged tankers have also caused disasters. Bodies of fresh water, and their associated ecosystems, have also been affected by human activity, from inadequate infrastructrure to deal with human waste to agricultural practices such as run-offs of artificial fertilizers. Some human rubbish, in the form of food waste, has enabled animals such as rats, foxes and even, in some places, polar bears, to thrive, surviving quite happily by going through our bins.

'Only within the moment of time represented by the present century has one species acquired significant power to alter the nature of the world.'

Rachel Carson, *Silent Spring* (1962), the first book to highlight the impact that chemical
pesticides were having on the natural world

Many non-human species have found their environments transformed by human activity, and this, together with hunting, has incurred a high rate of extinctions in recent centuries. At the same time, some non-human species have thrived on our intervention, notably those 'alien' species that humans have spread around the world, whether for utility or for ornament, without foreseeing the consequences. Cane toads, rabbits, rhododendrons, the kudzu vine, grey squirrels – these and many more have unbalanced their host ecosystems.

The role of humans in the bigger history of our planet is highlighted by the fact that more and more scientists now suggest that, in the context of deep time, we are in a new geological epoch. They dub this the Anthropocene (from Greek *anthropo-*, 'human' and *-cene*, 'new') epoch, an age in which human activities are having a global effect on the planet's ecosystems and geology. Some date this epoch to the beginnings of agriculture many thousands of years ago; some suggest it began with the Industrial Revolution, some 200 years ago; while others suggest a much more recent starting point: the first test explosion of an atomic bomb, on 16 July 1945.

THE FUTURE OF HUMANITY

Humans cannot help but wonder what the future holds, whether for themselves or their children's children. Religions around the world have come up with varied answers. Hinduism sees existence as extending indefinitely over vast cycles of time, in which individual souls are endlessly reincarnated. Other faiths, such as Christianity, envisage an apocalyptic end to human life on Earth within a definite (but unknowable) period or time, followed by a timeless eternity, in which the good are taken up to heaven and the wicked cast down to hell.

Today, however, many people are as likely to look to the predictions and models of scientists for glimpses of the likely future of our species, and our planet. The debate over climate warming and its likely impact on life on Earth has helped to focus minds. In November 2015, the World Meteorological Organization indicated that the year would be the hottest on record, with manmade emissions the chief cause. The average global temperature in February 2016 was higher by 1.35°C than that month's average for 1951–1980. Part of the rise in 2015 may have been caused by El Niño, a natural and mobile weather pattern marked by warming sea-surface temperatures in the Pacific; indeed the El Niño that year was one of the strongest on record.

The many conflicting interests of states and populations around the world make it hard to reach international agreements, not only on climate change, but also on other important issues. Rising population levels have created environmental pressures in many regions. Competition for resources such as oil has already led to extensive conflicts. Although the burning of fossil fuels contributes to global warming, oil still underpins large sectors of the global economy, and also the way of life of millions of people. Unless more effort is put into developing new and more sustainable sources of energy, the eventual depletion of oil reserves may well have an enormously disruptive impact on the way many of us live.

In the future, other natural resources, such as fresh water, are likely to lead to conflict. In the Middle East, for example, the rivers Tigris and Euphrates are vital sources of water for Iraq and Syria, but the amount they channel has been limited by the construction upstream in Turkey of dams for irrigation and hydroelectricity. The countries concerned have yet to come to an agreement over water sharing. Within countries, too, rising water consumption has depleted natural aquifers. In India, the boom in rice cultivation in the Punjab from the 1960s resulted in a serious drop in the water table. Extractions from the water table by manufacturing industry is a growing issue both in India and elsewhere. In 1990s Australia, the use of irrigation for cotton and rice in the Murray–Darling basin led to the movement of salt to the surface and to major losses of cultivable land. Something similar had occurred millennia before in ancient Mesopotamia.

'All I know about the future is that it is what you make of it.'

Walter Mosley, US novelist (1998)

Unless humans cooperate more effectively on the global scale, competition and conflict over diminishing resources are likely to lead to increasing disorder. If global warming remains unchecked, many parts of the world will end up uninhabitable, for example because of flooding or desertification. This would lead to great flows of refugees. It may be better to deal with the causes before we have to deal with the consequences.

If global warming continues, there is a possibility that agricultural productivity could collapse, and the resulting food shortages could spell the end of human life on the planet. However, there are other less predictable ways in which the human species might perish. Some of the more extreme scenarios include large-scale nuclear war, which could wipe out many in a moment. If any life forms managed to survive the blasts and the radiation, they might not be able to live through the ensuing nuclear winter: with so much debris thrown up into the atmosphere, the Sun's light could be blocked out for years, killing off the plant life at the foot of most food chains. A similar nuclear winter could be caused by the impact of a large meteor or a comet, or by the eruption of a supervolcano like the Yellowstone Caldera.

At the more bizarre end of the spectrum, science fiction has often imagined our world being invaded by hostile alien creatures. However, in spite of our much improved ability to observe other planets, and the discovery of water elsewhere in our own solar system, this remains a rather remote prospect. There is nonetheless the possibility of an enemy closer to home: a disease pandemic of such severity that the human species is wiped out. Some culprits for such a possible extermination include new strains of influenza, the Ebola virus, the return of the Black Death, a variation on the HIV/AIDS virus, extreme drug-resistant tuberculosis, a leak from biological warfare stockpiles, or perhaps another disease which we have not yet even heard of.

Many species of animal have already become extinct, and it is entirely possible that it will be the turn of our species to face extinction in the future. Even if we do survive that long, in a billion years or so, the Sun will become so hot that water on Earth will not stay liquid. As a result, all life on Earth will be extinguished.

Long after that, about 5 billion years from now, the Sun – like other stars of its size – will grow massively larger and become what astronomers call a red giant. This expansion will engulf all the inner planets, including the Earth.

> **'Earth felt the wound, and Nature from her seat**
> **Sighing through all her works gave signs of woe,**
> **That all was lost.'**
>
> John Milton, *Paradise Lost*, Book IX (1667)

We are in the early years of space travel, and it is uncertain whether we will ever be able to voyage between solar systems. However, when the Sun starts to get significantly hotter and expand, that may be the only way – if we haven't already become extinct – that human life can continue thereafter.

THE FATE OF THE UNIVERSE

How will the universe end? One possibility is that it will eventually suffer an implosion, as time, light and space collapse. In this theory, the expansion of the universe that started with the Big Bang will run out of momentum and the universe will start to contract back into itself.

This process has been called the Big Crunch, and this may in turn lead to another Big Bang, perhaps only the latest in an unimaginably long cycle. Alternatively, if there is not enough matter in the universe for gravity to bring about a Big Crunch, it has been conjectured that entropy might lead to the 'heat death of the universe', with all energy dissipated and the cosmos left for ever cold and lifeless.

More recently, in response to a range of observations of gravitational effects, cosmologists have proposed that over five-sixths of the mass of the universe may consist of something they call 'dark matter' – matter that has mass, but which cannot be observed using current technology. This would make the mass of the universe far greater than hitherto thought. It would also make the Big Crunch the more likely scenario.

Against this, observations of distant supernovae indicate not only that the furthest parts of the universe are moving away from us (as the Big Bang theory would expect), but that the rate of acceleration is increasing. Previously, physicists would have expected the momentum of the original expansion to slow down. Nothing has explained this rising acceleration, and scientists have posited a mysterious 'dark energy', acting counter to gravity, which might lead to the universe expanding indefinitely. This in turn has led to the suggestion of another possible way the universe might end – in a 'Big Rip', in which all objects in the universe no matter how large or small eventually disintegrate into elementary particles and radiation.

There are some obvious philosophical difficulties in contemplating the end of the universe. Firstly, none of the present scenarios – no matter how careful the mathematical modelling – are provable. They depend on current observations, which have been revised before and will be revised again. Secondly, the very act of thinking about the universe is a difficult one for the human mind. The universe is 'everything there is', and it is all but impossible for us to truly imagine it not existing. It is conceivable that the whole search for a start and an end to the universe is a misleading metaphor based on our own experience of all creatures being born, living for a lifetime and then dying. It is even possible that there are many parallel universes, of which our own is just one. But it would be impossible for a conscious being in this universe to have any direct knowledge of any of these other alternative worlds.

One of the advantages of studying big history, as opposed to human history, is that it gives us a humbling insight into how contingent and fleeting our existence is. Each one of us exists for a tiny fraction of the time that humans have lived on this planet. And the human species has existed for only a tiny fraction of the life of our solar system. Our solar system has in turn existed for only a fraction of the life of the universe and only came about because of a particular set of forces and matter combining in a certain way in a chaotic earlier period in our galaxy.

It is something of an achievement that we can know as much as we do about the universe and the history of our planet, but there comes a point where we have no choice but to accept that we will never know the full story of the universe in which we live.

PICTURE ACKNOWLEDGEMENTS

The Big Bang (page 12): Illustration by David Woodroffe

Continental drift (page 18): Map by David Woodroffe

Prototypes (page 37): Illustrations by David Woodroffe

Brachiosaurus (page 40): Illustration by David Woodroffe

Early migrations (page 58): Map by David Woodroffe

Lion Man (page 74): © Heritage Image Partnership Ltd / Alamy

Chariot burial (page 87): DeAgostini / Getty Images

Porsmose Man (page 91): Lennart Larsen / National Museum of Denmark

Ring of Brodgar (page 93): © Martin McCarthy / iStock

The Silk Road and other historical trade routes (page 101): Map by David Woodroffe

Early urban civilizations (page 105): Map by David Woodroffe

Chinese bronze coin (page 109): Universal History Archive / UIG via Getty Images

Gold coins of Croesus (page 110): DeAgostini / Getty Images

Rama (page 126): Illustration from *Myths of the Hindus & Buddhists*, 1914

Genghis Khan (page 143): © Adwo / fotolia

European colonization of the Americas (page 160): Map by David Woodroffe

Wright Brothers' first flight (page 181): Wright Brothers Collection / Prints & Photographs Division / Library of Congress / LC-DIG-ppprs-00626

European overseas colonization, 1914 (page 183): Map by David Woodroffe

Venus of Brassempouy (page 195): © The Natural History Museum / Alamy

Albert Einstein (page 199): George Rinhart/Corbis via Getty Images

Hydrogen bomb 'Ivy Mike' (page 223): Photo: The Official CTBTO Photostream

The USSR and its communist allies in Europe, 1950 (page 227): Map by David Woodroffe

Jakarta landfill site (page 249): Jonathan McIntosh / CC BY 2.0 / commons. wikipedia.org

INDEX

INDEX

Page numbers in *italic* refer to maps and illustrations

T